BASIC PLUMBING

ILLUSTRATED

By the Editors of Sunset and Southern Living

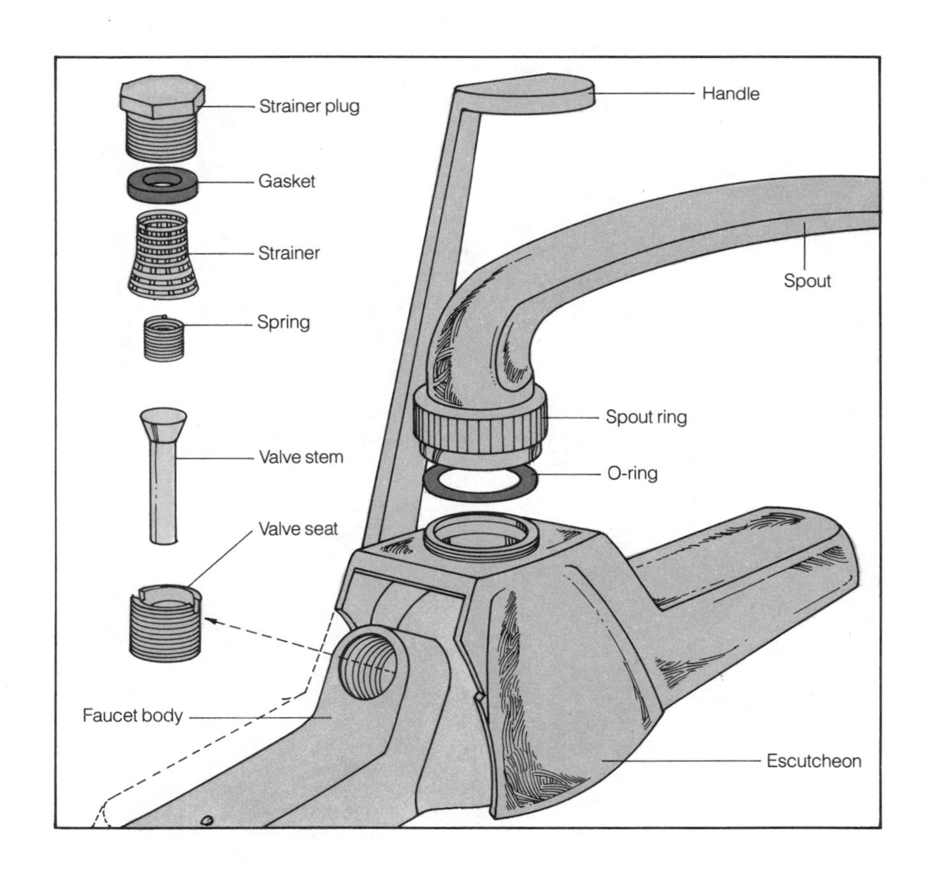

Book Editors
Lynne Gilberg
Don Vandervort

Coordinating Editor
Suzanne Normand Eyre

Design
Joe di Chiarro

Illustrations
Rik Olson
Mark Pechenik

A Plumbing Primer

Bothered by the incessant dripping of a leaking faucet? Eager to install an instant hot water dispenser in the kitchen? Anxious to save water in your garden? Take the mystery out of these and other plumbing projects with this complete illustrated guide to plumbing repairs and improvements.

The book begins at the beginning—with an explanation of how the plumbing system in a typical home works. It then takes you step by step through the most common household repair projects, from fixing faulty sink pop-ups to clearing clogged drains. The section on working with pipe prepares you for making your own plumbing improvements, whether you're simply replacing a faucet or putting in a new shower.

Finally, you'll learn how to install and maintain a garden irrigation system and how to conserve water both indoors and out.

Special thanks go to Karen A. L. Boswell, Michael Scofield, and Scott Atkinson for their editorial contributions, and to Donald E. Johnson, Building Official of Menlo Park, for his careful checking of the third edition.

Cover: Cover design by Williams & Ziller Design. Photography by Philip Harvey. Photo styling by JoAnn Masaoka Van Atta.

Editor: Elizabeth L. Hogan

Third printing January 1994

 Library of Congress Catalog Card Number: 92-81023. ISBN 0-376-01467-9 (SB); ISBN 0-376-09037-5 (SL). Lithographed in the United States.

CONTENTS

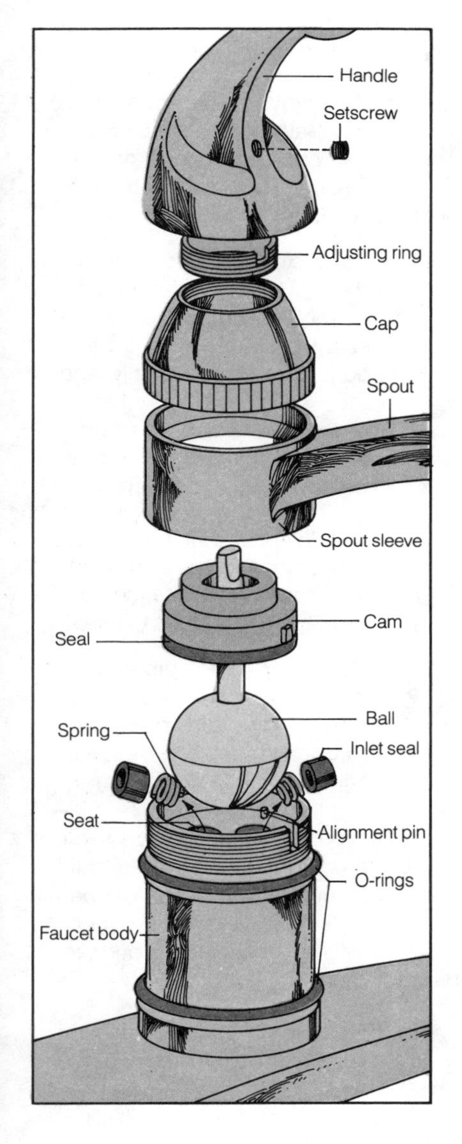

Features

Quick Answers to Plumbing Questions

Q **Sometimes when I flush the toilet it keeps running until I jiggle the handle. What should I do?**

A It sounds as if the toilet flush handle gets stuck, causing the tank stopper (see page 32) to stay open. This allows the water in the tank to continuously flow into the bowl. Make an adjustment by oiling, tightening, or replacing the flush handle. Turn to page 37 in the "Fix-it Handbook" for details.

Q **The hot water pipe to my kitchen sink is leaking. How do I make a quick repair?**

A To temporarily stop a small leak, break off a pencil point in the pipe hole and then wrap the pipe with three layers of plastic electrician's tape, extending 3 inches on either side of the hole. You can also clamp a piece of rubber, such as an old rubber glove, over the leak. For more on temporary patching, see page 43. If it's a major leak, though, turn off the water supply immediately at the main shutoff valve (see page 23) and replace the pipe (pages 47–58).

Q **When my washing machine shuts off abruptly, the water supply pipes make a loud banging noise. Is there anything I can do to stop this?**

A Water hammer (the noise you've described) occurs because the water in the pipes slams to a stop, causing a shock wave and a hammering noise. It's not only annoying but also destructive to the pipes.

You can minimize or eliminate water hammer by installing air chambers—dead-end pieces of pipe (see page 45). Most washing machine manufacturers recommend extra-long chambers—up to 24 inches—to provide added cushion for abrupt turn-offs. Page 83, "Connecting a Washing Machine," will help you with this and related problems.

To learn how to install or restore (drain water from) an air chamber, read "Noisy Pipes" on page 45.

Q **I have an old brass faucet in my bathroom that I want to keep, but it constantly drips hot water from its spout. I've replaced the washer in the hot water faucet, but the spout still drips. How can I repair the faucet so it stops wasting my hot water?**

A A spout leak in a compression faucet like yours is caused either by a defective seat washer or a damaged valve seat (see pages 13–14). You can replace most valve seats with exact duplicates, using a valve seat wrench (see "Tools to Keep Handy," page 11). If a worn valve seat can't be removed, use a valve seat dresser to grind it until smooth. For more on fixing compression and washerless faucets, see pages 13–18.

Q **Should we use plastic or copper pipe to install a new water softener in our system?**

A First check your local plumbing code on plastic pipe. Some areas don't allow it at all, and others allow it for everything but drinking water supply pipes. If permitted, plastic piping can be a good choice, because it's less expensive, easier to work with (to cut, join, and maneuver), self-insulating, and resistant to weather and corrosion. But be forewarned: Plastic pipe has its disadvantages, too. Before you make your decision, read pages 47–55 in "Pipefitting Know-how" for complete details. Also, see pages 88–89 on installing a water softener.

Q **What causes our water heater to make loud rumbling noises?**

A The two most common causes for such noises are steam and sediment in the tank (see "Water Heater Problems," pages 38–39). You can often correct steam problems by merely lowering the thermostat setting. If you suspect a faulty thermostat, turn the setting all the way down; then if the heat source doesn't go off, replace the thermostat.

To get rid of problem-causing sediment in your water heater, open the drain valve at the bottom of the tank and drain off a little water until it runs clear. Draining the sediment should eliminate noise problems and allow your heater to operate more efficiently. For details on replacing a water heater, turn to pages 84–85 in "Plumbing Improvements."

Q **We always have a soap ring in our bathroom sink because the pop-up stopper doesn't open far enough for the water to drain out quickly. How do I adjust it?**

A If the stopper is so tight that the sink doesn't drain properly, you'll need to get under the sink to reset the pivot rod (see page 21) by squeezing the spring clip and inserting the rod in the next higher hole. Also remove the pop-up stopper and clean it periodically. Hair and debris can cause sluggish drainage.

Q **I need to cut a new piece of copper pipe for the supply run to my sink. What are the best tools and techniques to use?**

A It's best to use a pipe cutter (see Drawing 6C on page 49) with a specially designed blade for copper pipe. You can also use a fine-toothed hacksaw, but making a straight cut with it is more difficult. After you've cut the pipe, clean off any burrs (inside or out) with a half-round file. For more details on copper pipe, including how to remove, measure, solder, and hang the pipe, see pages 52–55.

Q **Our friends have an instant hot water dispenser mounted on their kitchen sink. They say it conserves energy because it eliminates the need to boil water for tea, instant coffee, soup, and the like. Can I install one of these hot water dispensers myself?**

A You can do the plumbing and installation of most hot water dispensers in an afternoon, but unless you're familiar with wiring techniques, leave the electrical hookup to a professional. The project involves attaching the dispenser faucet onto the sink rim or countertop, tapping into the cold water pipe with a saddle tee fitting, and mounting the hot water holding tank under the sink. For installation details, turn to page 82 in "Plumbing Improvements."

By the way, your friends are right about the dispenser being an energy-saver. See the feature on page 59, "Energy conservation tips," for additional information on energy-saving appliances and devices.

Q I don't really understand what makes a plumbing system work. The purposes of pipes to supply water and to drain waste seem clear enough, but what's the purpose of vent pipes?

A Plumbing works because of constant water pressure (about 50 pounds per square inch) in hot and cold supply pipes, the pull of gravity in drainpipes, and the balance of air pressure in vent pipes. Each fixture needs a vent to get rid of sewer gas and prevent a buildup of pressure in the pipes. To learn more, see "How Plumbing Works," beginning on page 7.

Q We want to add a second sink in the master bathroom. Can we extend the pipes that are already there?

A Yes, you can. You'll need to tap into the existing supply, drain, and vent pipes, run new piping to the desired location, and hook up the new fixture. For step-by-step instructions, turn to the "Adding a second sink" feature on page 71 and to the information on installing sinks on pages 72–73.

Q I've tried using a plunger and chemical drain cleaners, but my tub is still clogged. What else can I do before I resort to calling a plumber?

A When working with water that contains chemical cleaners, use rubber gloves, bail out any standing water, don't plunge, and avoid splashing.

For a stubborn clog, use a snake as described on pages 27 and 29. Feed the snake down the drain or overflow pipe to the trap to break up the blockage. If that doesn't work, the problem is probably deep down in the main drain. For the next step, see pages 30–31 on main drain clogs.

Q Can you give me some advice before I replace a worn-out toilet with a new one?

A Since a toilet is a major water guzzler in a home, choose one that conserves water (see "Water-conserving Fixtures," pages 105–106). Buy a toilet that's ready to install, with flush assembly in place. Most important, carefully measure the roughing-in distance (see page 78) and select a new toilet that will fit properly in the space.

Actual installation takes a little muscle (lifting the old and new toilet off and on) and some time, but doing the job yourself can save you the high cost of a plumber.

Q How hard will it be to plan and install a sprinkler system for my front yard?

A It's easier than you may think. The chapter "Exterior Plumbing" on pages 90–101 will guide you through the entire process. You'll learn how to design the system, how to install all the necessary components, and how to maintain it in good working order. Although digging trenches can be tiring and hooking up valves and a timer can be tricky, the actual installation of pipes and sprinkler heads is simple, especially if you use plastic pipe. The same chapter also contains information on designing and installing tubing and emitters for a drip irrigation system, the most efficient method of watering plants.

Q When the weather gets hot and humid, our toilet tank sweats so much that the floor tiles below the tank are starting to loosen. What can I do to prevent toilet tank condensation?

A Before treating a sweating tank problem, be sure that a leak is not the culprit (see page 37). To stop condensation on the tank, install a foam jacket (sold in hardware stores) or pieces of ½-inch-thick foam rubber inside the tank (see page 37). You'll need to empty the water from the tank before you glue the foam in place.

Another solution—though more expensive and involved—is to install a tempering valve (see page 37).

Q The trap under my kitchen sink has corroded through and started to leak. I'd like to replace it but don't know where to start.

A The hardest part of replacing a trap is loosening the couplings that are sometimes frozen in place on the old trap. Start by emptying the trap through the cleanout plug (if it has one) into a pail. Use a tape-wrapped wrench and counterclockwise force to remove the couplings at the tailpiece and drainpipe. Before installing a new trap, coat the threads of the connecting pipes with pipe joint compound or pipe-wrap tape to guard against leaks. "Trap Problems," pages 24–25, shows how.

Q The water doesn't drain out of our dishwasher. Any suggestions before I call for repair?

A There are three common causes for standing water in the bottom of a dishwasher: a plugged strainer basket in the tub of the dishwasher, a dirty air gap, or a dirty hose loop that vents the appliance. Each is easily remedied—just clean out dirt, grease, or food buildup. To learn how, see the dishwasher chart on page 41.

Q We're considering adding an entirely new bathroom to our house. Can you let us know what to expect once we get to the plumbing part of this extensive project?

A Read "Roughing-in & Extending Pipe" on pages 60–65 before you start. This section outlines the planning process, code restrictions, venting options, and techniques for locating and exposing pipes, making pipe connections, running new pipe, and roughing-in fixtures.

You'll Need to Know

How Plumbing Works

If your plumbing experience has been limited to turning a faucet on and off, you'll be surprised at the simplicity of the system of pipes behind that faucet. Actually, there are three systems: supply, drain-waste, and vent. Before you begin any plumbing project, large or small, it's a good idea to be familiar with these systems. Once you understand how plumbing works, you'll find that making repairs or adding fixtures is simply a series of logical connections.

- *The supply system* carries water from underground water mains, a storage tank, or a well into your house and around to all the fixtures (sinks, showers, toilets) and to such appliances as the washing machine and dishwasher.
- *The drain-waste system* carries used water and waste out of the house into a sewer or a septic tank.
- *The vent system* carries away sewer gases and maintains atmospheric pressure inside the drainpipes, preventing deadly gases from entering your home.

The supply system

If you get your water from a water company, it's probably delivered by underground water main through a water meter and a main shutoff valve. You'll find the meter either in your basement or crawl space (to guard against freezing in cold-winter areas) or outdoors, near your property line. The main shutoff valve—which turns the water for the whole house on and off—is usually situated near the water meter.

If your water is provided by a water company but is not metered, the main shutoff valve is likely to be at your property line; if you can't locate it, check with the water company. Where water comes from a private well, the shutoff valve is usually located where the water supply line enters the house, or at the wellhead—or both.

As a rule, the water supply pipe that enters the house is a 1-inch-diameter pipe that's under about 50 pounds of pressure per square inch (psi), though pressures vary in different localities. (For a discussion of water pressure problems, see page 42.)

If the water in your house is softened or filtered, the treatment units will be attached near the point where the water enters the house. Usually, a water softener is placed on a branch pipe that goes to the water heater, so that only the hot water used for laundry or bathing is softened. Chemical injectors or filters will be on the main pipe in the basement or crawl space.

Once inside the house, the supply pipe branches out into pipes of smaller diameters to deliver water to all fixtures and water-using appliances. A separate horizontal pipe makes a stop at the water heater, then runs parallel to the cold water pipes to the kitchen, laundry, and bathrooms throughout the house (see **Drawing 1**).

Hot and cold water pipes are usually ¾ inch in diameter. Branches that feed the fixtures are generally ½-inch galvanized iron, copper, or plastic pipe. Local codes, practices, and the age of your house will affect the kinds of pipe and fittings you'll find and will determine what you can use if you're planning any changes or additions (see "Pipefitting Know-how," beginning on page 46).

Pipes that run vertically from one floor to the next are called risers. Long risers are often supported at their bases by platforms and anchored to wall studs. Horizontal pipe runs are generally secured to floor joists.

Supply pipes are installed with a slight pitch in the runs, sloping back toward the lowest point in the system so that all pipes can be drained. Sometimes at the lowest point there's a valve that can be opened to drain the system.

Shutting off the water supply. Most fixtures and water-using appliances have their own shutoff valves to enable you to work at one place without cutting off the water supply for the entire house. To be prepared for an emergency, everyone in the household should learn how to turn off the water supply, both at individual fixtures and at the main shutoff valve (see page 23).

Gas and heating system pipes. If you're planning to do a plumbing job yourself, you must be able to distinguish the water supply pipes from pipes that carry natural gas or propane into your home for a gas stove, dryer, or water heater. A gas pipe is usually a black or galvanized pipe that runs from the gas meter directly to a gas appliance or heating system. A separate shutoff valve for emergencies is required on each supply pipe. Don't try to work on gas piping—call a professional.

Heating system pipes demand equal caution. To locate your heating pipes (hot water or steam), trace them between each heating outlet and the furnace or other heat source. And by all means, leave repairs to an expert.

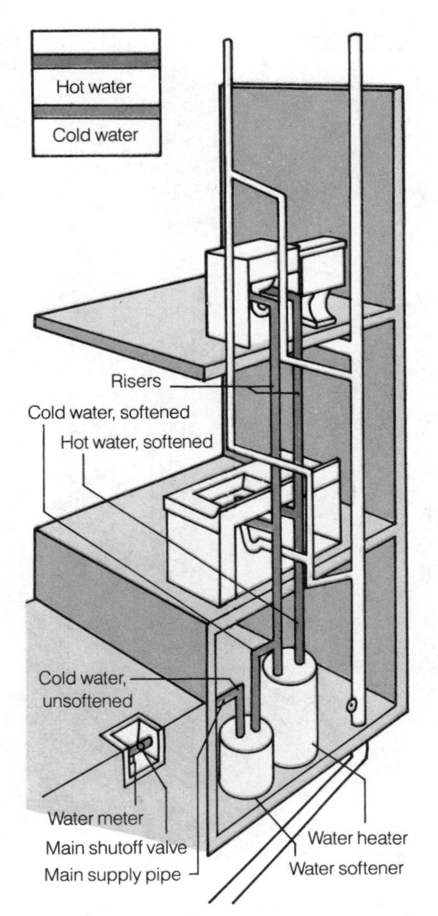

Drawing 1. Supply system

. . . How Plumbing Works

The drain-waste system

Unlike the supply system, which brings in water under pressure, the drain-waste system gets rid of water and wastes by the force of gravity. These pipes lead away from all fixtures at a carefully calculated slope. If the slope is too steep, water will run off too fast, leaving particles behind; if it's not steep enough, water and waste will drain too slowly and may back up into the fixture. The normal pitch is ¼ inch for every horizontal foot of pipe.

The workhorse in the drain-waste system is the soil stack, a vertical section of 3 or 4-inch-diameter pipe that carries waste away from toilets (and often from other fixtures) and connects with the main house drain in the basement or crawl space. From here, the wastes flow to a sewer or septic tank (see **Drawing 2**).

Since any system will clog occasionally, cleanouts are placed in the drainpipes. Ideally, there should be one cleanout in each horizontal section of drainpipe, plus an outdoor cleanout to give access to the sewer or septic tank. A cleanout is usually a 45° Y fitting (see **Drawing 3**) or 90° T fitting (see **Drawing 4**), that makes it possible for a snake or auger to be inserted into the main drainpipe to clear away obstructions.

The vent system

To prevent dangerous sewer gases from entering the home, each fixture must have a trap in its drainpipe and must be vented. A trap is a bend of pipe that remains filled with water at all times to keep gases from coming up the drains. The two main types are P traps (see **Drawing 5**) and, in older homes, S traps (see **Drawing 6**); the latter are no longer allowed by many plumbing codes.

The vents in the drain-waste system are designed to get rid of sewer gas and to prevent pressure buildups in the pipes. The vents come off the drainpipes downstream from the traps and go out through the roof (see **Drawing 2**). This maintains atmospheric pressure in the pipes and prevents siphoning of water from the traps.

Each plumbing fixture in the house must be vented. Usually, a house has a 4-inch-diameter main vent stack with 1½ to 2-inch vent pipes connecting to it. In many homes—especially single-story homes—widely separated fixtures make it impractical to use a single main vent stack. In this situation, each fixture or fixture group has its own waste connection and its own secondary vent stack.

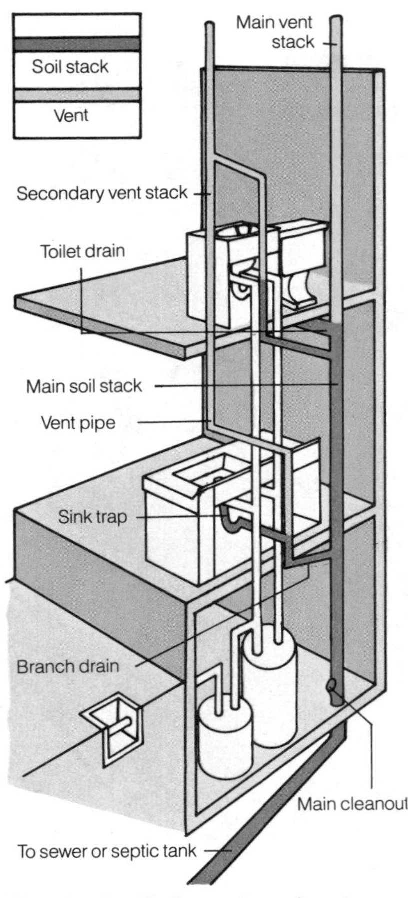

Drawing 2. Drain-waste and vent (DWV) systems

Drawing 3. Y fitting with cleanout insert

Drawing 4. T fitting

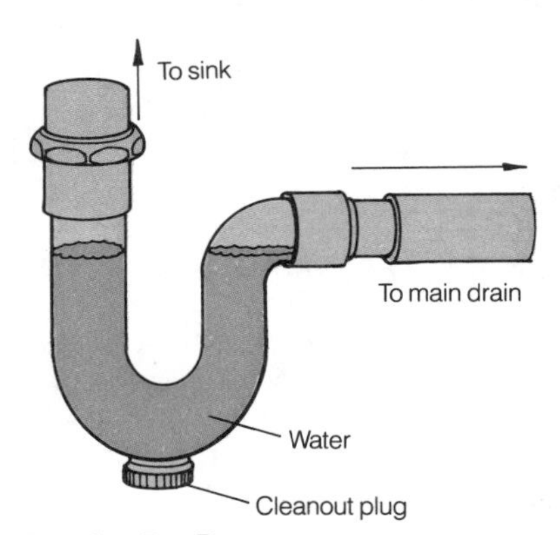

Drawing 5. P trap

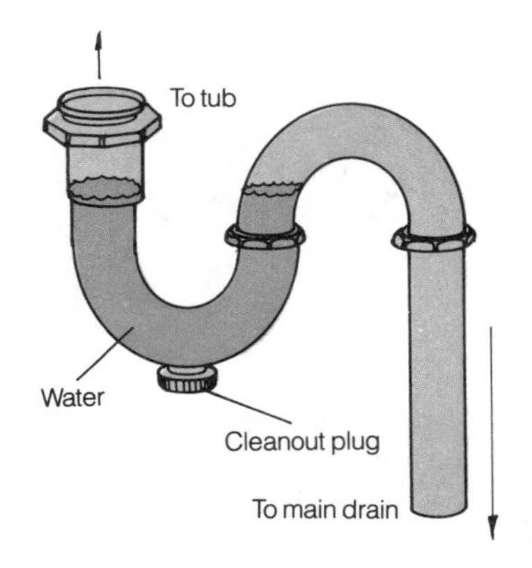

Drawing 6. S trap

A Typical Plumbing System

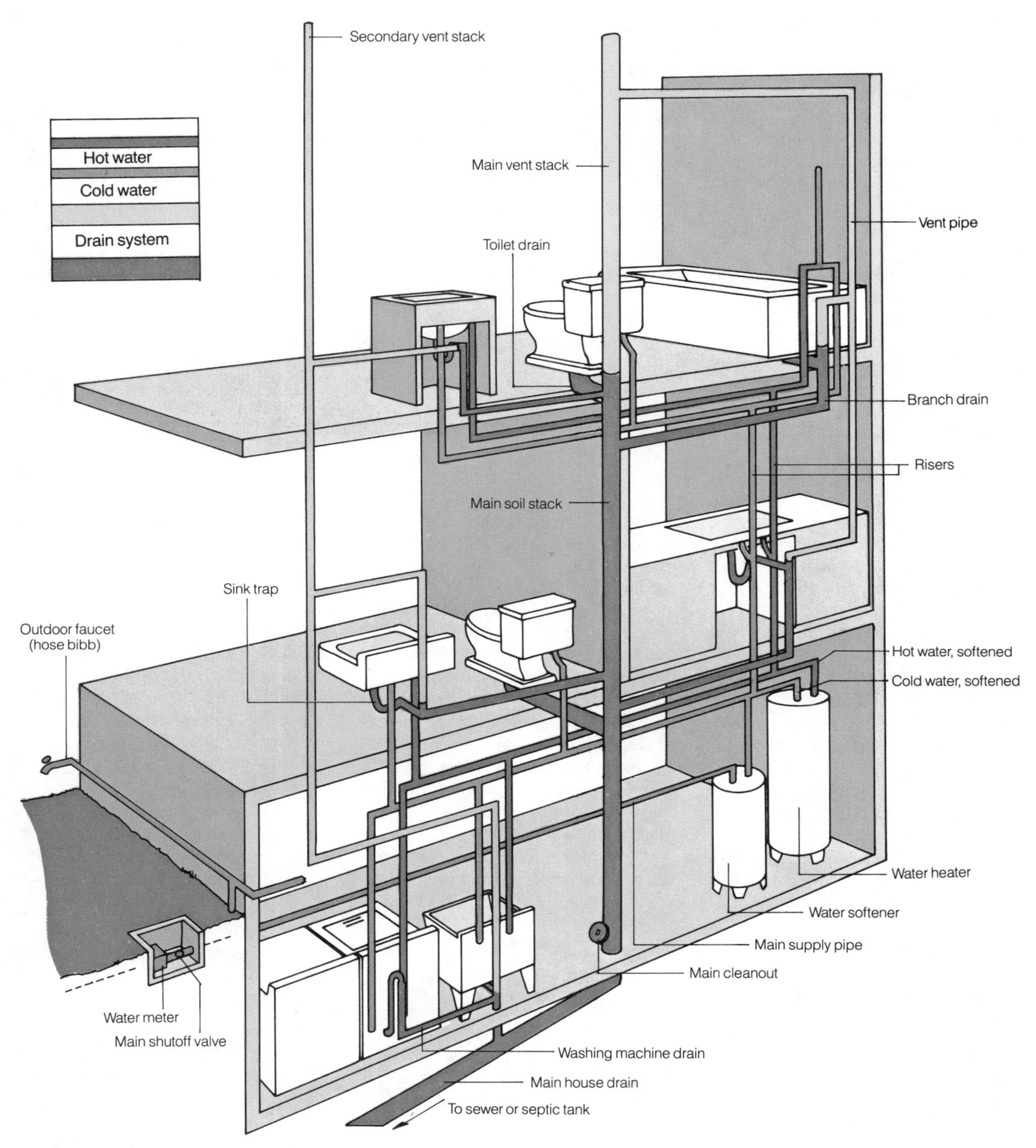

Three separate but interdependent systems—supply, drain-waste, and vent—make up a typical plumbing system. Here's a behind-the-scenes look at each. The supply system brings water (under pressure) to your house, through a water meter (if you have one) and a main shutoff valve. The supply pipes branch off to all fixtures and water-using appliances inside and to faucets outside the house. The drain-waste system carries used water and waste out of the house (with the help of gravity) via the main soil stack to the sewer or septic tank. The vent system gets rid of sewer gases for each fixture and maintains proper pressure inside the drainpipes.

Plumbing Codes & Permits

Why do we need plumbing codes and permits? To establish standards to protect the health and safety of the community, and to see that those standards are followed. Faulty plumbing can cause serious health and safety hazards, such as toxic gas backups, bursting pipes, floods, and electrical shorts.

There are six model plumbing codes in print, but regulations regarding methods, materials, and design differ from one state, county, or municipality to the next. Local codes supersede the model codes.

Materials and methods specified in local codes are constantly being updated. The most controversial material in plumbing today is plastic pipe. It's prohibited entirely in some areas of the country; in others it's permitted for drain-waste and vent pipes only.

Just like a contractor, a do-it-yourself plumber must abide by all the rules and regulations of the codes. If your work violates a code, you run the risk of having to rip it all out.

Before you begin any work, be sure your plumbing plans conform to local codes and ordinances. Discuss your plans in detail with a local building inspector, and be sure the methods and materials you're planning to use are acceptable. The inspector will tell you whether or not you need a plumbing or building permit.

Projects that involve changes or additions to your plumbing system—specifically, to the pipes—usually require permits. You won't need a permit, though, for replacements—a new fixture or appliance—or for emergency repairs, as long as the work doesn't alter the plumbing system. When in doubt, be sure to check.

How to read your water meter

Reading a water meter isn't nearly as complicated as all those dials make it seem, and it can help you detect leaks as well as keep track of water usage.

Three types of meters. Your home's water usage is probably measured by one of the three common meters.

The *six-dial meter* is by far the most prevalent for residential use (see **Drawing A**). Five of its six dials (labeled 10, 100, 1,000, 10,000, and 100,000 for the number of cubic feet of water they record per revolution) are divided into tenths. The needles of the 10,000 and 100 meters move clockwise, and those of the other three move counterclockwise. The remaining dial, which is usually undivided, measures a single cubic foot per revolution.

To read the six-dial meter, begin with the dial labeled 100,000, noting the smaller of the two numbers nearest the needle. Then read the dial labeled 10,000 and so on.

The *five-dial meter* is read in exactly the same way as the six-dial meter, except that single cubic feet are measured by a large needle that sweeps over the entire face of the meter (see **Drawing B**).

The *digital read-out meter* looks like an odometer, giving you the total number of cubic feet at a glance (see **Drawing C**). This meter may also have a small dial that measures a single cubic foot per revolution.

Measuring usage and detecting leaks. Keeping track of the water used for a specific task or by a specific appliance is as simple as reading your meter. Just subtract the "before" reading from the "after" reading.

To track down a possible leak, turn off all the water outlets in the house and note the position of the one-foot dial on your meter. After 30 minutes, take another look at the dial. If the needle has moved, you have a leak.

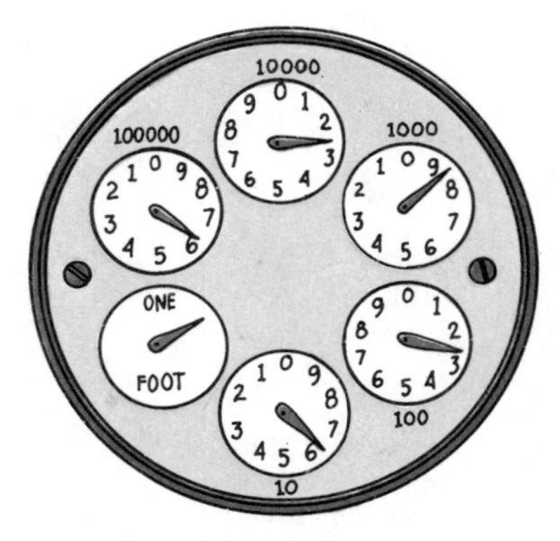

Drawing A. Six-dial meter reads 628,260.

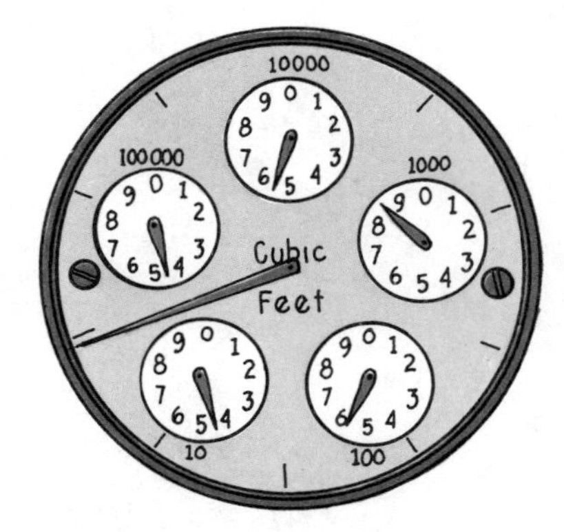

Drawing B. Five-dial meter reads 458,540.

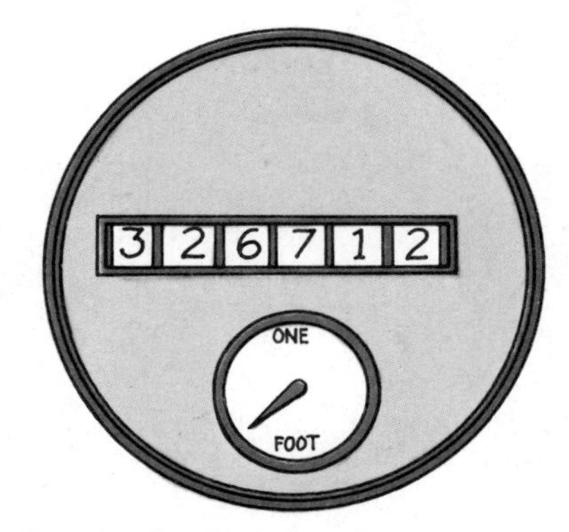

Drawing C. Digital read-out meter reads 326,712.

Tools to Keep Handy

To do any job right, you need the right tools and materials—and plumbing is certainly no exception. Shown here are most of the basics, other than everyday items such as a flashlight.

Here's a list of some miscellaneous items you'll want to keep on hand to hold things together in an emergency: sheets of foam rubber, a length of old hose, automotive hose clamps, assorted faucet washers, O-rings, wire coat hangers, and a variety of nuts, bolts, and metal washers. Specialized tools for particular types of pipe are shown on pages 46–65.

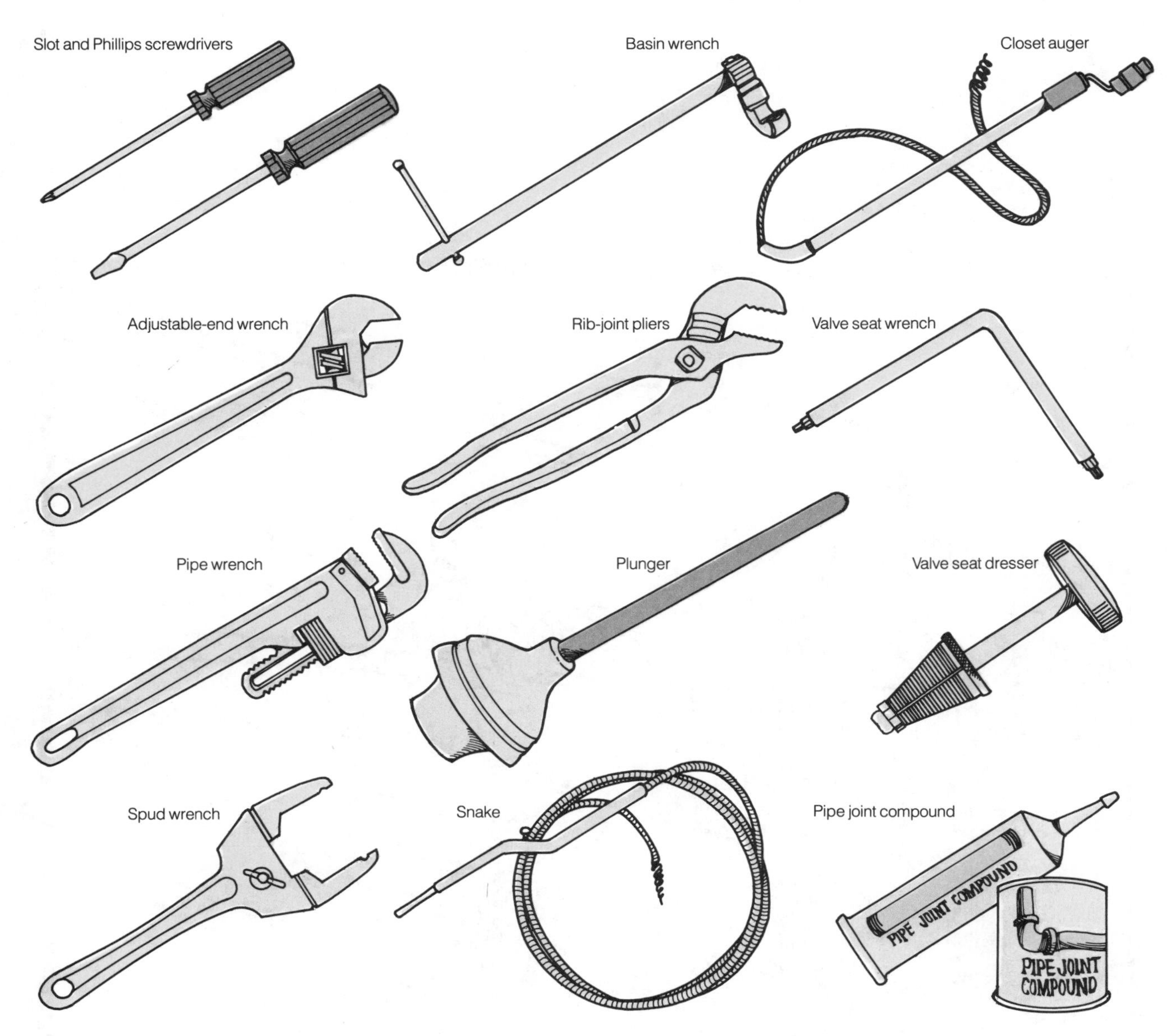

Slot and Phillips screwdrivers, common household tools, are essential for fixing leaking faucets and making other repairs. **Adjustable-end wrench** has smooth jaws made to fit small nuts, bolts, and square and hexagonal fittings. **Pipe wrench**, or monkey wrench, has toothed jaws designed to grip pipe. **Spud wrench**, wide and toothless, adjusts to fit large nuts on toilets and sinks. **Basin wrench** allows easy access to nuts behind sinks and other hard-to-reach places. **Rib-joint pliers**, or slip-jaw pliers, open wide enough to remove drain traps. **Plunger**, also called a plumber's helper, dislodges clogs using alternate pressure and suction. Funnel-cup type shown is designed for toilets; it folds flat for drains. **Snake**, or drain-and-trap auger, stretches 10 to 20 feet to remove deep blockages in a drain. **Closet auger**, a 3 to 6-foot tool with a crank handle, unclogs toilets; it works like a snake but has a protective housing to prevent scratching the bowl. **Valve seat wrench**, with one end square and the other hexagonal, removes worn or damaged valve seats. **Valve seat dresser** grinds and smooths faulty nonreplaceable valve seats in old faucets. **Pipe joint compound**, available in cans, tubes, and sticks, is used to lubricate, seal, and protect pipe threads when pipe and fittings are being assembled.

Fix-it Handbook

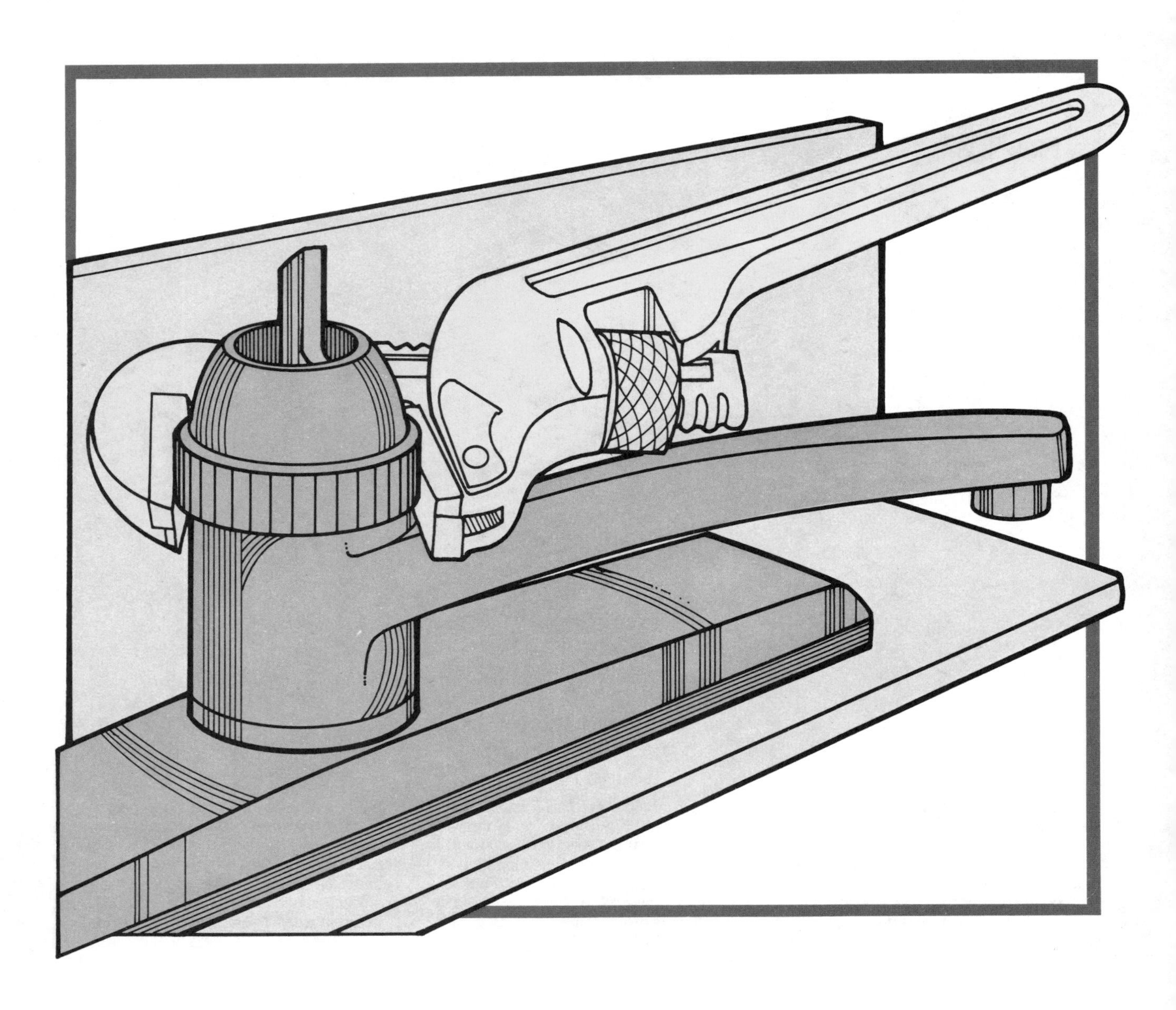

Leaking Compression Faucets

That constant drip, drip, drip from a spout or persistent oozing from a handle not only means wasted water—money down the drain—but also becomes annoying enough to demand attention.

The first step in fixing the faucet is identifying the type you're dealing with. There are two basic types of faucets. One is a long-standing design—usually with two handles and one spout—called a compression, stem, or washer faucet (see **Drawing 1**). The other is a more recent type called a washerless or noncompression faucet (pages 15–18)—in it, the mix of hot and cold water is controlled by one handle or knob.

The compression faucet closes by a screw pushing and compressing a washer against a valve seat. Ideally, when the faucet is turned off, the stem is screwed all the way down and the washer fits snugly into the valve seat, stopping the flow of water.

Cap
Handle screw
Handle
Stem nut
Packing nut
Stem
Packing
Threads
Faucet body
Seat washer
Washer screw
Valve seat

Drawing 1. Compression faucet comes apart easily. Handle-type faucets may have stems threaded in opposite directions.

Before starting any faucet repair, plug the sink drain so that small parts can't fall down it, and line the sink with a towel to prevent damage from parts or tools accidentally dropped. Line up disassembled pieces so that you'll be able to put them back together in the right order.

CAUTION: Before doing any work, turn off the water at the fixture shutoff valves or at the main shutoff valve (page 23). Open the faucet.

Handle leaks

If the leak comes from around the handle stem, try tightening the packing nut slightly. If the leak persists, you'll have to replace the packing inside the nut.

To take the faucet apart (see **Drawing 2**), remove the decorative cap on top of the handle, using a blunt knife or screwdriver. Undo the handle screw and pull the handle straight up off its stem. Remove the stem nut and packing nut with an adjustable-end wrench. Using the handle, turn the stem beyond its fully open position to unscrew it; then lift it out.

The packing on the faucet stem is either a rubber O-ring, a packing washer, or graphite twine (see **Drawing 3**). If it's an O-ring (A), pinch the old one off with your fingers and roll on a new one that's exactly the same size and type. If it's a packing washer (B), remove the old one and push an exact replacement onto the stem. If it's graphite-impregnated twine (C), scrape away all of the old material and wrap new twine clockwise—five or six times—around the faucet stem.

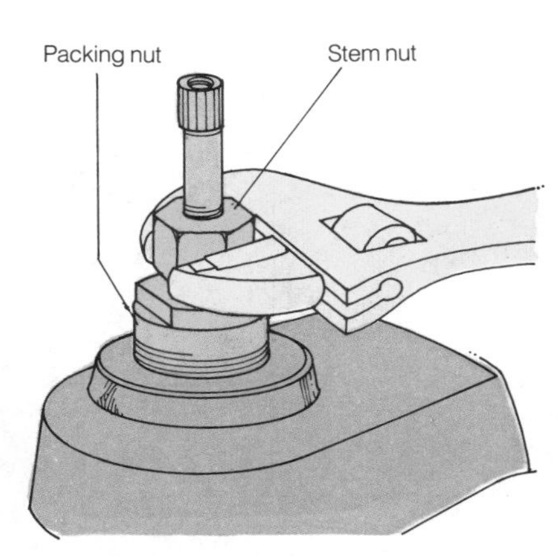

Drawing 2. To take apart a compression faucet, remove the nuts. Turn the stem beyond fully open and lift out.

Before replacing the packing nut, lubricate the threads of the stem and nut with petroleum jelly. Tighten the packing nut and replace the handle.

Spout leaks

Even though it's the faucet spout that's dripping, it's one of the handles that needs repair. To discover which handle needs work, turn off the shutoff valves under the fixture one at a time. The leak will stop when one or the other of them is turned off, and you will have narrowed down the problem.

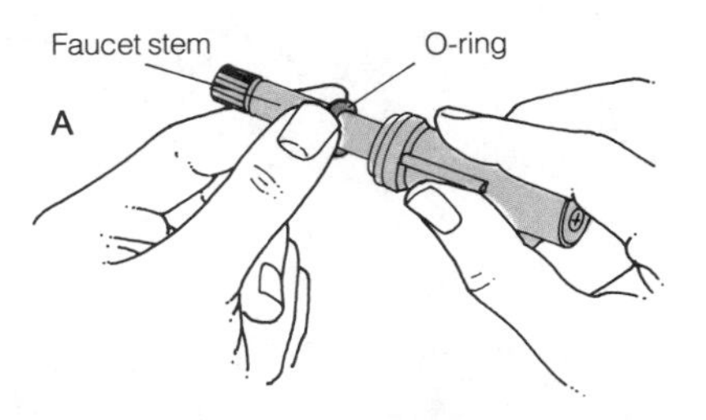

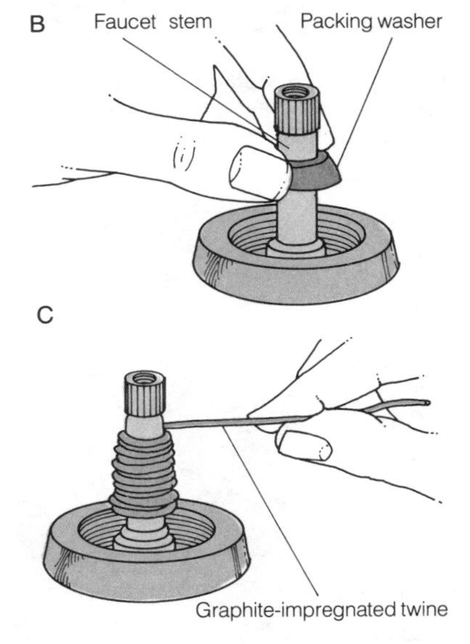

Drawing 3. To replace the packing, take off an O-ring (A), packing washer (B), or graphite-impregnated twine (C).

. . . Leaking Compression Faucets

A spout leak is usually the result of a deteriorated washer or valve seat in the faucet. Remove the handle and stem. At the bottom of the stem is a brass washer screw or seat screw that goes through the center of a rubber-like seat washer (see **Drawing 4**). If the washer is cracked, grooved, or marred, carefully remove the screw and replace the washer with a new, identical one. Seat washers may be flat or beveled (conical). If the washer is beveled, be sure the beveled edge faces the screw head when you install it on the stem.

During replacement, the shank of the washer screw may break off. If it does, you can replace it with a new swivel-head washer (see **Drawing 5**) or replace the entire faucet stem.

A swivel-head washer has two prongs that snap into the bottom of the stem to compress against the valve seat. To install, drill a hole in the end of the stem to receive the prongs.

If the washer isn't the trouble-maker, look deeper. A damaged valve seat could be causing the leak. A pitted, corroded, or gouged valve seat prevents the seat washer from fitting properly. Fortunately, most compression faucets have a replaceable valve seat (see **Drawing 6**).

You'll need a valve seat wrench (see page 11) to make the exchange. Insert the wrench into the faucet body and turn it counterclockwise to remove the seat. At a plumbing supply store, buy an exact duplicate of the seat. Before installing the new seat, lubricate its threads with pipe joint compound.

If the faulty valve seat can't be removed (it may be built into the faucet), use a simple, inexpensive tool called a valve seat dresser (see **Drawing 7**). It grinds down any burrs on the seat, making it level and smooth. Buy the largest dresser that will fit the faucet body.

Insert the valve seat dresser until the cutter sits on the valve seat. Turn the tool handle clockwise until the seat is smooth. Remove the metal filings with a damp cloth.

After the new seat washer or valve seat is in place, be sure to lubricate the threads of the stem with plumber's grease. Then you can put the handle back.

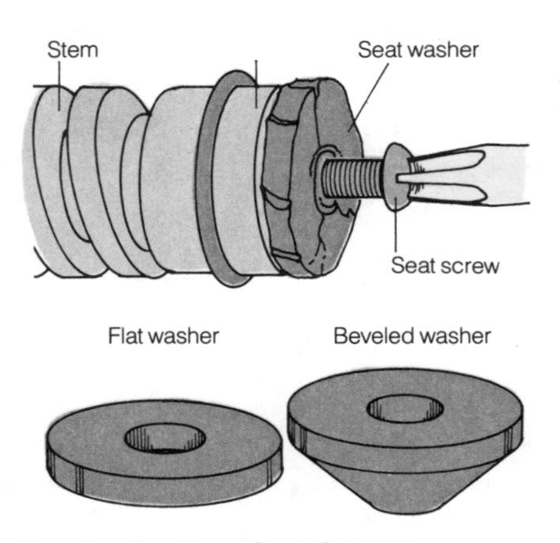

Drawing 4. To replace the seat washer—flat or beveled—remove the screw and install a new washer.

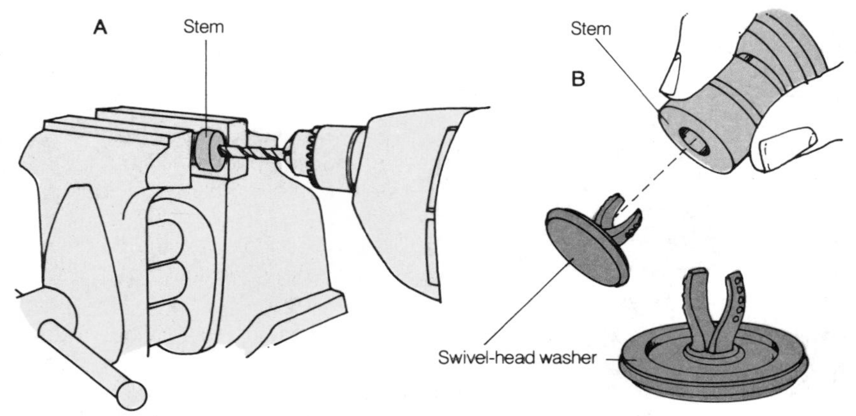

Drawing 5. To install a swivel-head washer, drill a hole in the bottom of the faucet stem (A) and snap the washer into the hole (B).

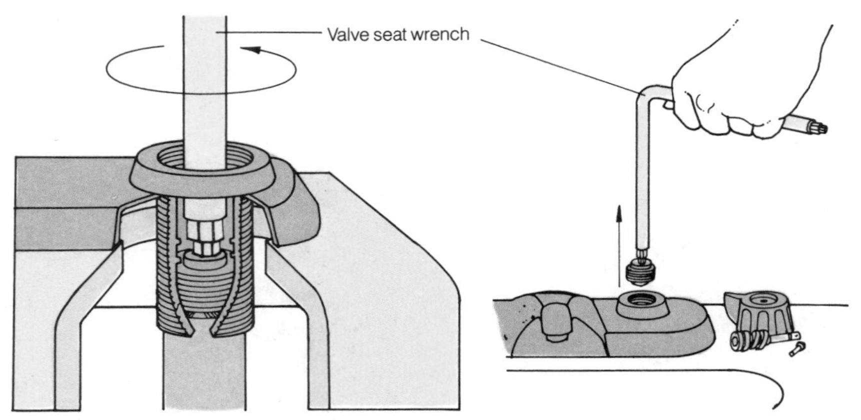

Drawing 6. To remove the valve seat, insert a valve seat wrench into the faucet body and turn counterclockwise.

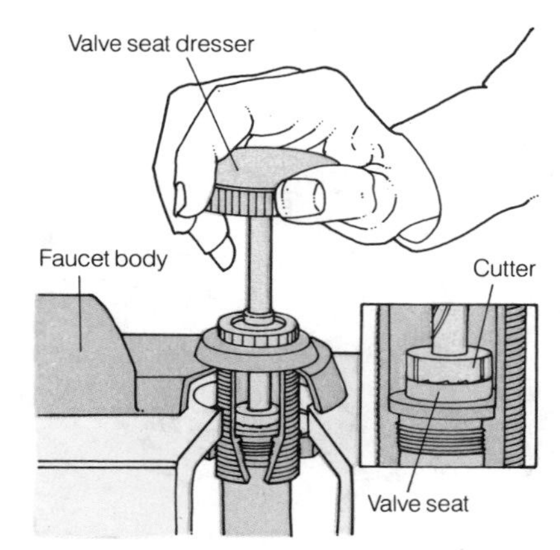

Drawing 7. To use a valve seat dresser, place the tool in the valve seat and turn until the seat is smooth.

Leaking Washerless Faucets

A washerless faucet usually has a single lever or knob that controls the flow and mix of hot and cold water by aligning interior openings with the water inlets. Washerless faucets generally work for years without fail, but when one needs repair you must replace some or all of the mixture and flow parts. How you'll do this depends on the type of faucet—disc, valve, ball, or cartridge (see **Drawing 8**).

When taking a washerless faucet apart, look for screws and nuts in odd places, such as under handles or at the base of the spout. Once you can see inside the faucet, you can determine what kind of control the unit has.

Disc faucets

This type of washerless faucet (see **Drawing 9**) relies on two discs that connect with the handle to mix hot and cold water. The disc assembly seldom wears out. More often, a rubber inlet seal proves to be the Achilles' heel.

CAUTION: Before doing any work, turn off the water at the fixture shutoff valves or the main shutoff valve (page 23). Open the faucet to drain the pipes.

To repair a leak at the base of a disc faucet, remove the setscrew under the faucet handle and lift off the handle and decorative escutcheon. Then remove the cartridge (see **Drawing 10**) by loosening the two screws that hold the cartridge to the faucet body.

Under the cartridge, you'll find a set of inlet seals (see **Drawing 11**). Take each one out and replace any worn ones with exact duplicates. Also check for sediment buildup around the inlet holes; scrape away any deposits to clear the restriction. When reassembling the faucet, be sure to align the inlet holes of the cartridge with those in the base of the faucet.

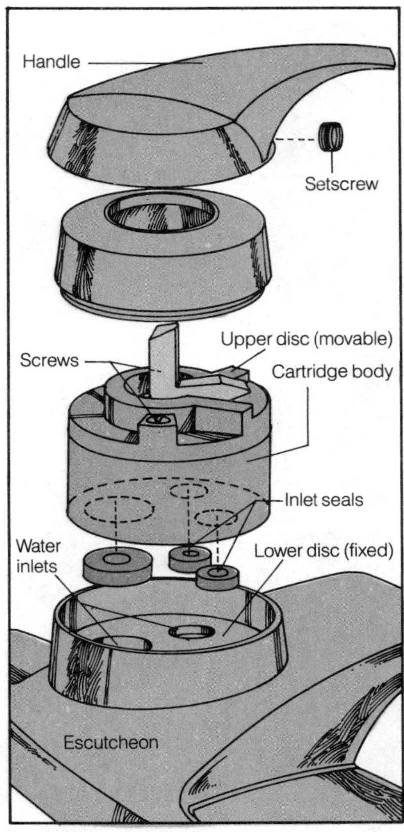

Drawing 9. Disc faucet

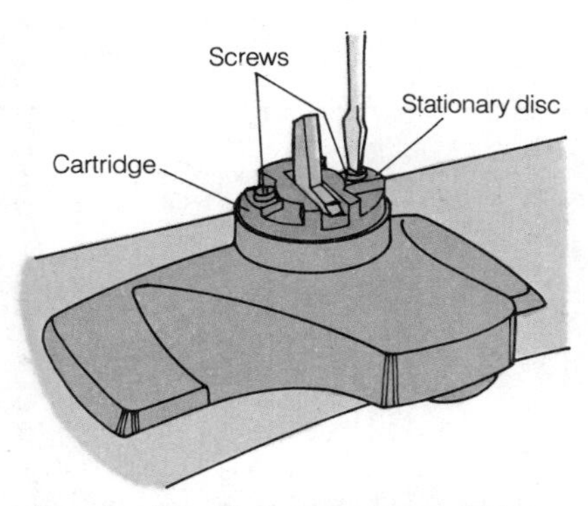

Drawing 10. To remove the cartridge, loosen the screws that hold the cartridge to the faucet body.

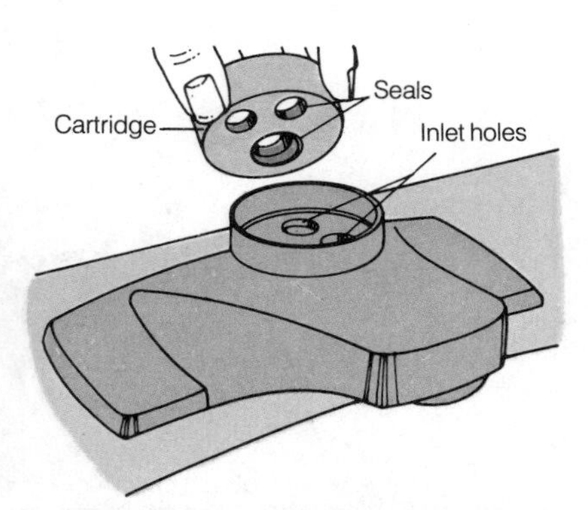

Drawing 11. Check the inlet seals for wear, and scrape away any sediment in the inlet holes.

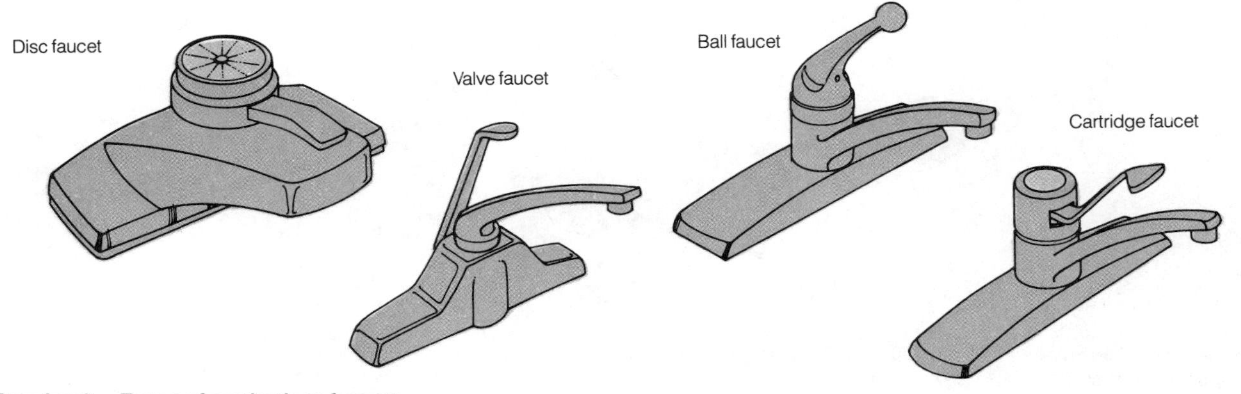

Drawing 8. Types of washerless faucets

. . . Leaking Washerless Faucets

Valve faucets

This type of faucet (see **Drawing 12**) is so called because it has a pair of valve stem assemblies (one for hot water, one for cold) through which water flows up and out the spout. Moving the handle forward and backward controls the flow; moving it from side to side controls mix. All the parts are replaceable; though these faucets are no longer made, you can buy a kit that has all parts (except strainers and plugs) at a plumbing supply store, or you can buy the parts separately.

CAUTION: Before doing any work, turn off the water at the fixture shutoff valves or the main shutoff valve (page 23). Open the faucet to drain the pipes.

Using a pipe wrench wrapped with electrician's tape, turn the spout ring counterclockwise.

If your only problem is a leak at the base of the spout, replace the O-ring (see **Drawing 13**) with an identical one and reassemble the faucet.

If the spout drips, you'll need to replace one of the valve parts or the valve seat (see **Drawing 14**). Remove the spout and escutcheon. Unscrew the hexagonal strainer plugs on either side of the faucet and take out the valve parts—a gasket, strainer, spring, and valve stem—by hand. Use a valve seat wrench to remove the valve seat. Replace any worn or corroded part. Lubricate the threads of the valve seat with plumber's grease and reassemble the faucet.

If flow from a valve-type faucet is sluggish, it's likely that the strainers are clogged with sediment from hard water. Clean the parts with an old toothbrush and soapy water; rinse each thoroughly and reinstall.

If the handle of a valve faucet has loosened, first tighten the screw that holds the handle to the cam assembly (see **Drawing 15**). If the handle still wobbles after you tighten the handle screw, remove the screw. Chances are that the unthreaded portion of the screw beneath the screw head is worn flat. Solution: Replace the screw.

If none of these remedies works, tighten the adjusting screw atop the cam assembly about one quarter turn.

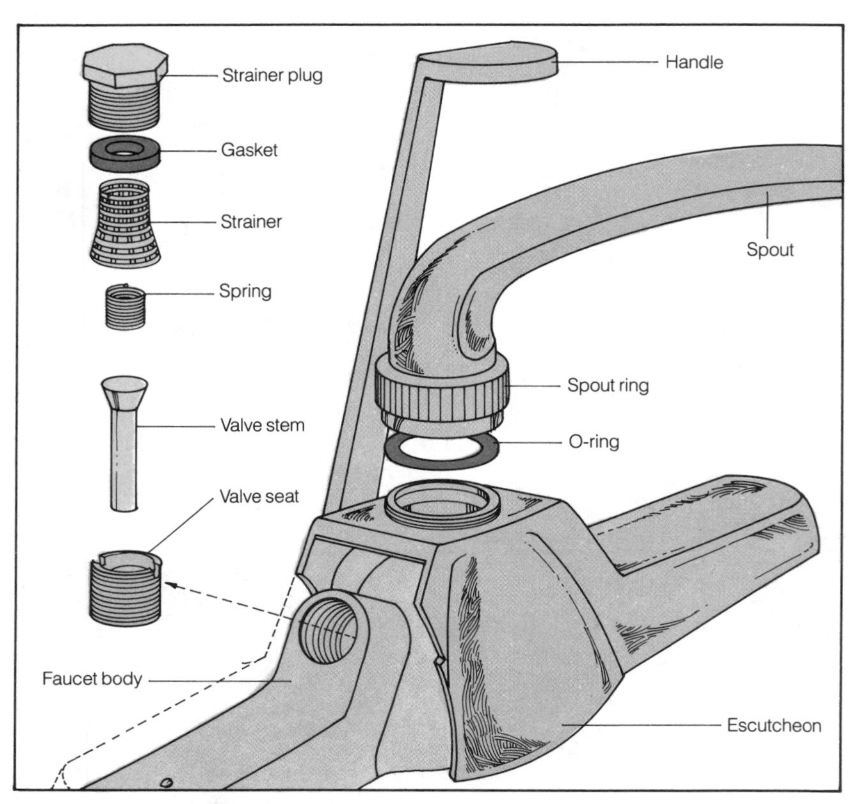

Drawing 12. Valve faucet

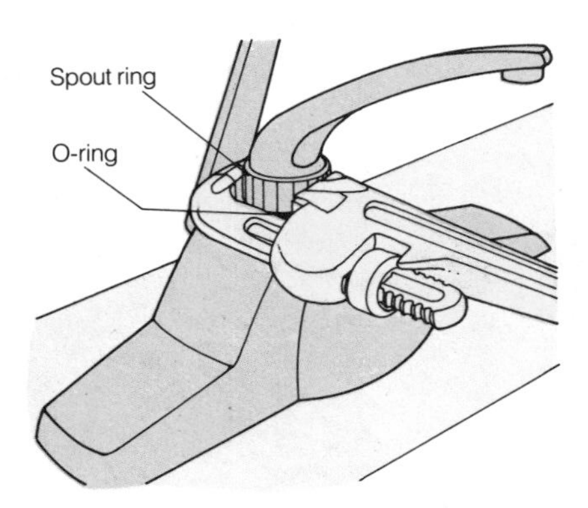

Drawing 13. To replace an O-ring, turn the spout ring counterclockwise, using a tape-wrapped pipe wrench.

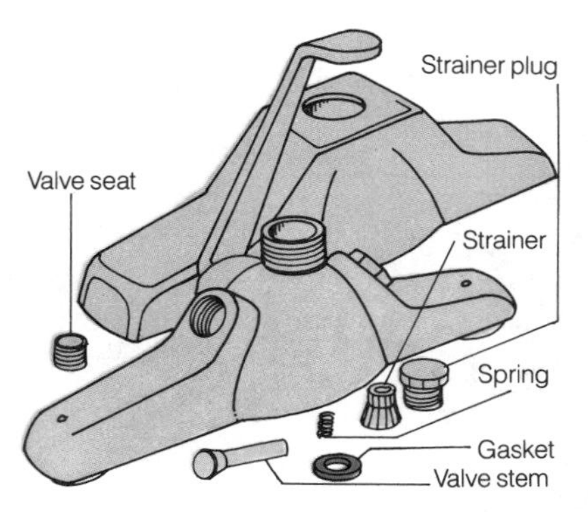

Drawing 14. To remove the valve assembly, unscrew the strainer plugs and take out the valve parts by hand.

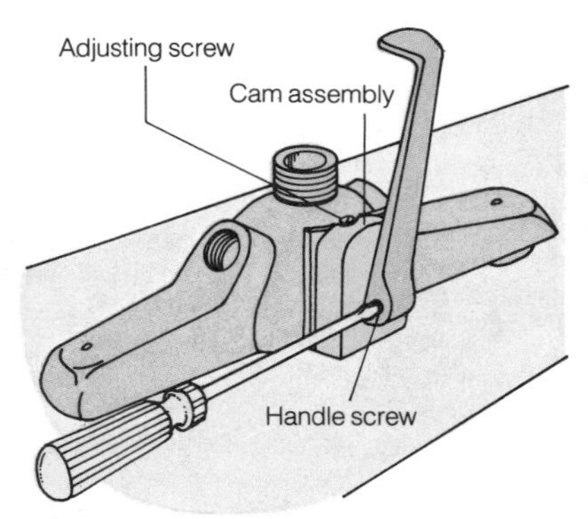

Drawing 15. To fix a loose handle, tighten the screw at the base of the handle or turn the adjusting screw on top of the cam.

Ball faucets

Inside every ball faucet (see **Drawing 16**) is a slotted metal ball atop two spring-loaded rubber inlet seals. Water flows when the openings in the rotating ball align with hot and cold water inlets in the faucet body.

If the handle of a ball faucet leaks, tighten the adjusting ring or replace the seal above the ball. If the spout of a ball faucet drips, the inlet seals or springs may be worn and need replacement. If the leak is under the spout, you must replace the O-rings or the ball itself.

CAUTION: Before doing any work, turn off the water at the fixture shutoff valves or the main shutoff valve (page 23). Open the faucet to drain the pipes.

Remove the faucet handle by loosening the setscrew with an Allen wrench. Use tape-wrapped rib-joint pliers to unscrew the cap (see **Drawing 17**).

Lift out the ball-and-cam assembly. Underneath are two inlet seals on springs. Remove the spout sleeve to expose the faucet body.

To replace the seals and springs (see **Drawing 18**), use needle-nose pliers to lift out the old parts. With a stiff brush or penknife, remove any buildup in the inlet holes. If new O-rings are needed, apply a thin coat of petroleum jelly to them to stop leaks at the base of the faucet.

Before reassembling the faucet, check the ball; if it's corroded, replace it. To reinstall the ball-and-cam assembly (see **Drawing 19**), carefully line up the slot in the ball with the metal alignment pin in the faucet body. Also be sure to fit the lug on the cam into the notch in the faucet body.

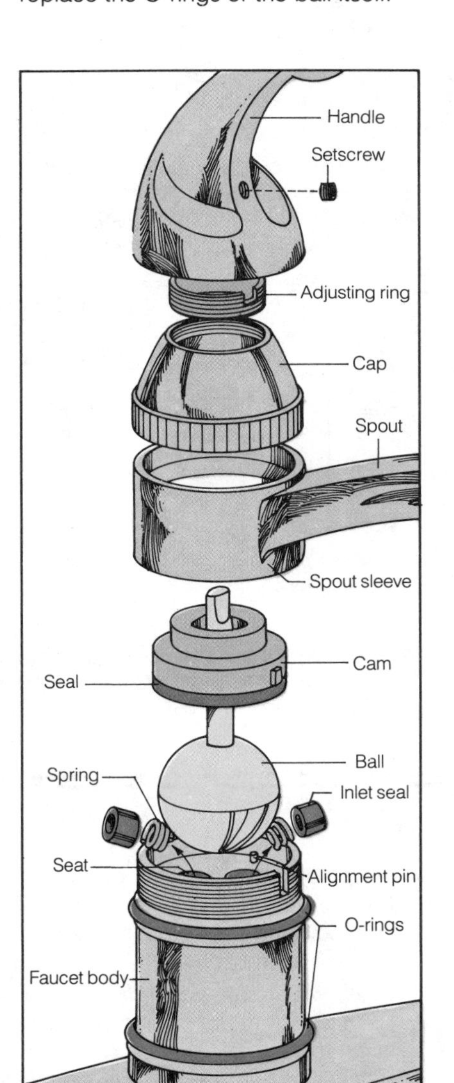

Drawing 16. Ball faucet

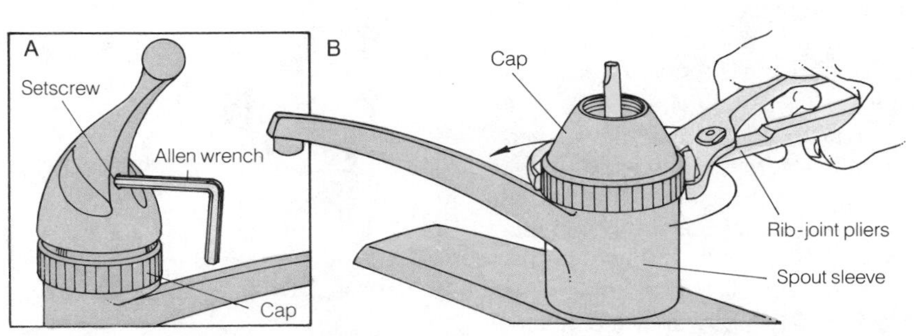

Drawing 17. To take apart a ball faucet, remove the setscrew with an Allen wrench (A), and the cap with rib-joint pliers (B).

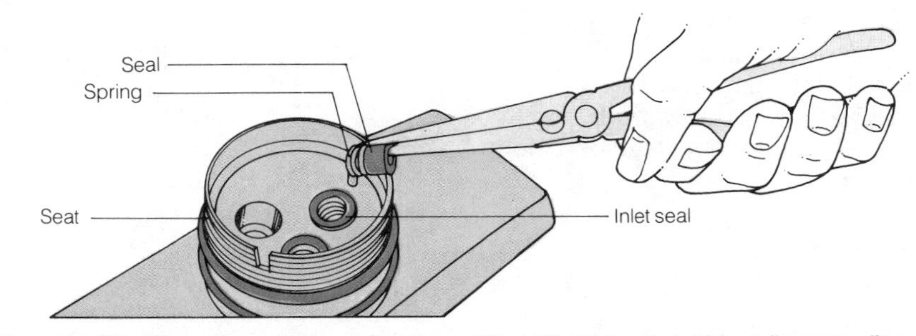

Drawing 18. To replace seals and springs, lift out the old parts with needle-nose pliers and replace with exact duplicates.

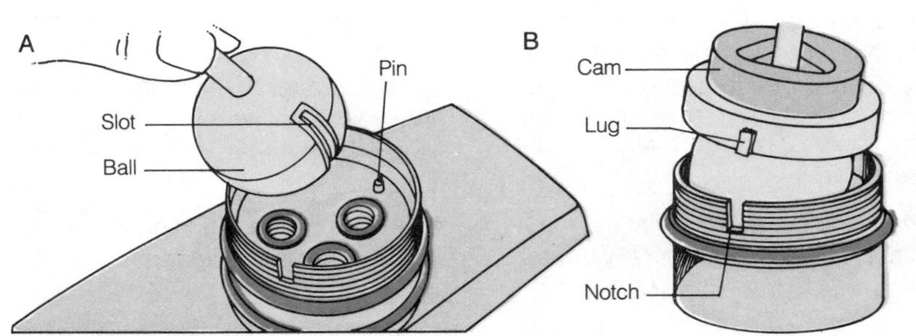

Drawing 19. To reinstall the ball-and-cam assembly, align slot in ball with pin (A), and fit lug on cam into faucet body notch (B).

. . . Leaking Washerless Faucets

Cartridge faucets

These washerless faucets (see **Drawing 20**) have a series of holes in the stem-and-cartridge assembly that align to control the mixture and flow of water. Usually problems with this type of faucet occur because the O-rings or the cartridge itself must be replaced.

The cartridge is held in place by a retainer clip, which may be on the inside or on the outside of the faucet. Once you remove the retainer clip, the stem-and-cartridge assembly simply lifts out.

CAUTION: Before doing any work, turn off the water at the fixture shutoff valves or the main shutoff valve (page 23). Open the faucet to drain the pipes.

Take the faucet apart by removing the handle screw and the cap atop the faucet. Moving the spout sleeve back and forth, gently pull it off the faucet body. Then lift off the retainer ring.

Next, remove the cartridge (see **Drawing 21**). You'll find the retainer clip just under the rim of the faucet body. Using a screwdriver or needle-nose pliers, remove the clip from its slot. Grip the stem of the cartridge with pliers and lift it out. Examine the O-rings on the cartridge and replace them if they show signs of wear. On swivel-spout models, apply petroleum jelly to the new O-rings before installing.

If the O-rings are in good shape, it's the cartridge assembly that has seen its day. Take the old one to the plumbing supply store and buy an exact duplicate.

Installing a cartridge (see **Drawing 22**) is a simple task, but remember to read the manufacturer's instructions first. Cartridges vary; the most common type has a flat side that must face front—otherwise your hot and cold water supply will be reversed. Also, be sure to fit the retainer clip snugly into its slot.

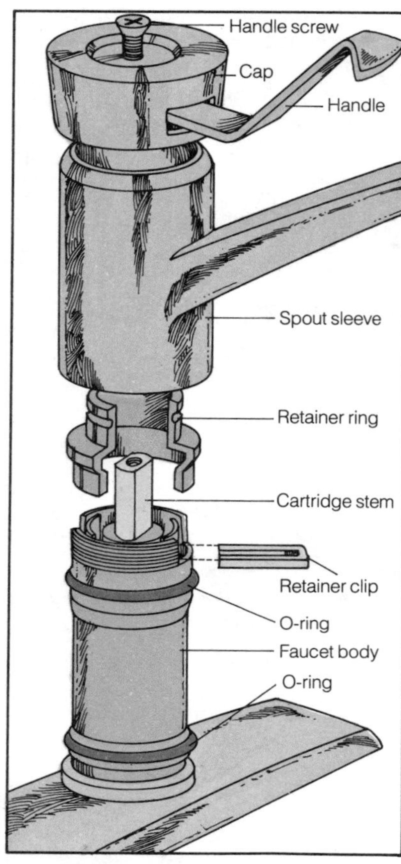

Drawing 20. Cartridge faucet

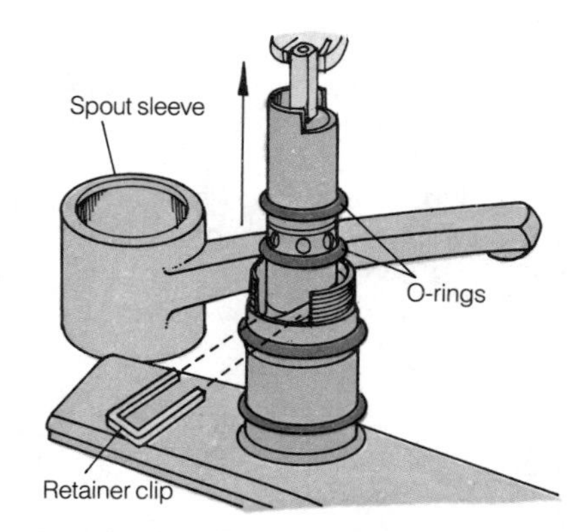

Drawing 21. To take apart a cartridge faucet, remove the spout sleeve and retainer clip, and lift out the cartridge.

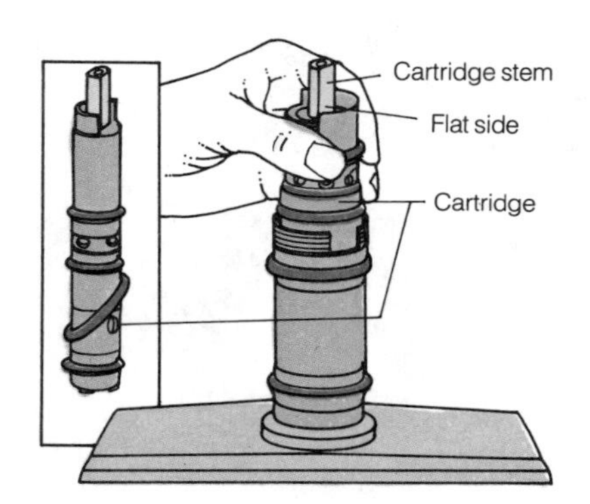

Drawing 22. To install a cartridge, face the flat side (if there is one) forward and fit the retainer clip into its slot.

Minimizing overflow damage

If an appliance overflows, a pipe bursts, or some other plumbing emergency occurs, you'll quickly learn that water on the loose can wreak havoc. To minimize damage, follow these guidelines:

- *Turn off the water supply* at the main shutoff valve (see page 23) before taking time to trace the source.
- *If an appliance is the culprit,* turn off all the power by throwing the main circuit breaker or the main switch so that you're not working with electricity and water—a dangerous combination.
- *If a pipe is leaking,* spread waterproof dropcloths and pans in the dripping area.
- *If a pipe has burst,* dam doorways with rolled-up rugs or blankets to help keep water from spreading throughout the house.
- *Use a water-safe vacuum* or rent a submersible-motor pump to remove large volumes of water.
- *Call the fire department* if the flooding is extensive.

Aerator, Sink Spray & Diverter Problems

An aerator, sink spray, or diverter valve can cause a full range of problems quite apart from common faucet failures. But most of these repairs are quick and easy—as long as you take care to properly reassemble any array of intricate parts.

Almost every faucet has, at the tip of its spout, an aerator that mixes air and water for a smooth flow. You should clean aerators periodically to remove mineral and debris buildup.

Sink sprays also have nozzle aerators that can clog, causing the diverter valve to malfunction. Other sink spray maladies can be cured with new washers or a new hose. If problems persist, look to the diverter valve—inside the base of the faucet. A possible solution: the diverter valve, which causes water to be rerouted from the spout to the sink spray hose, may need to be cleaned or replaced.

CAUTION: Before doing any work, turn off the water at the fixture shutoff valves or the main shutoff valve (page 23). Open the faucet to drain the pipes.

Cleaning an aerator. Unscrew the aerator from the end of the spout (see **Drawing 23**) or spray nozzle (see **Drawing 24**). Disassemble and set the parts aside—in order—for easy reassembly. Clean the screens and disc with a brush and soapy water; use a pin or toothpick to open any clogged holes in the disc. Replace worn parts and flush all parts with water before putting them back together.

Repairing a spray hose. If the hose leaks at the spray head, unscrew the head from the coupling at its base. Separate the hose from the coupling by snapping off the retaining ring (see **Drawing 25**). If the hose washer under the coupling is damaged, replace it; then flush out the hose.

If the hose leaks at the base of the faucet spout, undo the coupling under the sink (see **Drawing 26**), using rib-joint pliers or a basin wrench (see page 11). Getting at the coupling can be awkward; you'll need to lie on your back under the sink. Once the coupling is detached, inspect the entire length of hose for kinks or cracks. If you find defects, replace the hose with a new one of the same diameter—nylon-reinforced vinyl seems most durable.

Cleaning a diverter valve. You'll need to take off the faucet spout to get at the diverter (see pages 13–18 for help). Once you have access to the inside of the faucet body, loosen the screw atop the diverter valve just enough to lift the valve from the seat (see **Drawing 27**). Take the valve apart and clean its outlets and surfaces with an old toothbrush and water.

If cleaning the diverter, aerator, and spray hose doesn't improve the sink spray's performance, replace the diverter valve with an exact duplicate.

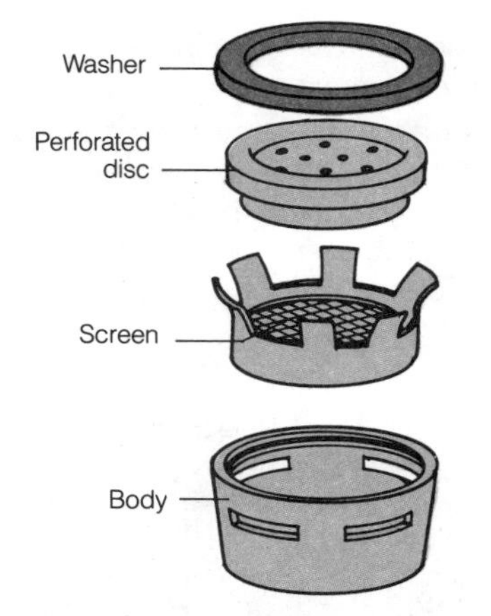

Drawing 23. Aerator for faucet spout

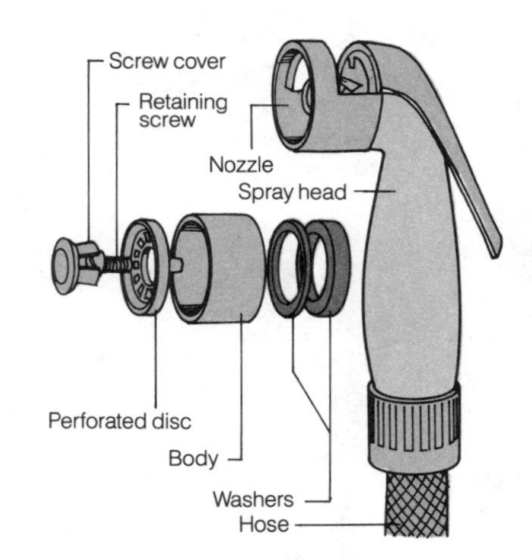

Drawing 24. Aerator for sink spray nozzle

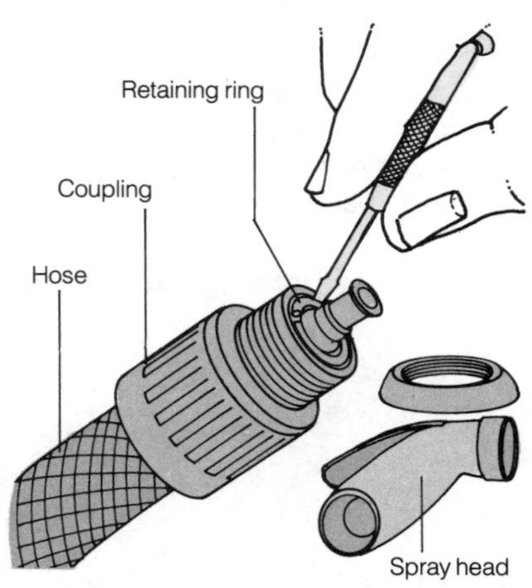

Drawing 25. To remove a spray head, unscrew the coupling at the base of the head and snap off the retainer ring.

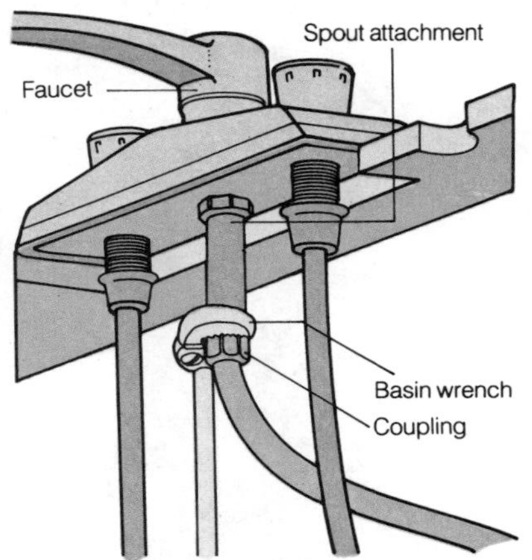

Drawing 26. To replace a spray hose at the spout, use a basin wrench to unfasten the coupling under the sink.

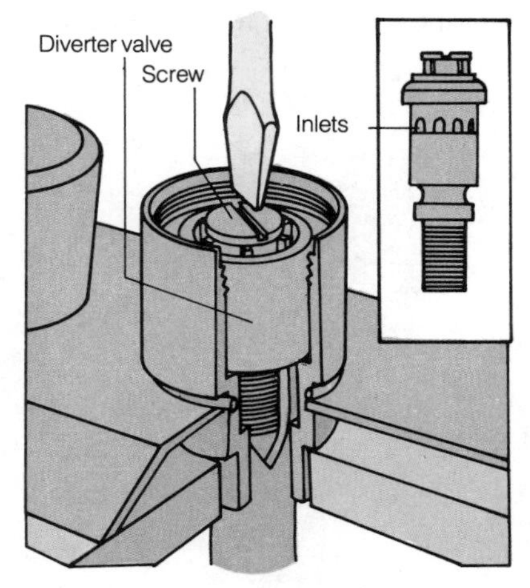

Drawing 27. To remove a diverter valve, loosen the screw atop the valve enough to lift the valve from the seat.

Leaking Sink Strainers

If your sink doesn't hold water, or if there's a leak under the sink, it's time to examine the strainer—the part that opens and closes the drain and traps large food particles.

There are two types of sink strainers (see **Drawing 28**): one held by a locknut, the other by a retainer and three screws.

For a locknut strainer (see **Drawing 29**), tap with a screwdriver and hammer to loosen the lugs on the locknut. Be careful not to damage the sink. Remove the locknut, metal washer, and rubber gasket from the bottom of the strainer body. Lift the strainer from the sink.

For a retainer-type strainer, simply undo the three screws on the retainer and disassemble as above.

Thoroughly clean the area around the drain opening. Check the rubber gasket and metal washer for wear; get exact replacements if needed. Apply a ⅛-inch-thick bead of plumber's putty around the underlip of the strainer body (see **Drawing 30**) and insert it in the opening. Press down firmly for a tight seal between the sink and the strainer.

If the strainer is held in place by a locknut, work from beneath the sink to place the rubber gasket and metal washer onto the strainer body, and hand tighten. Have a helper hold the strainer from above to keep it from turning while you snug up the locknut with a spud wrench. If you're working alone, place the handles of a pair of pliers in the strainer and hold a screwdriver between the handles for counterforce while you tighten the nut (see **Drawing 31**). Replace the coupling and connect it to the tailpiece. Wipe any excess putty from the sink's surface.

If the strainer is held in place by a retainer, it's an easy one-person installation. With the gasket and washer in place, fit the retainer over the strainer body and evenly tighten the three screws until secure. Screw on the coupling and check for leaks.

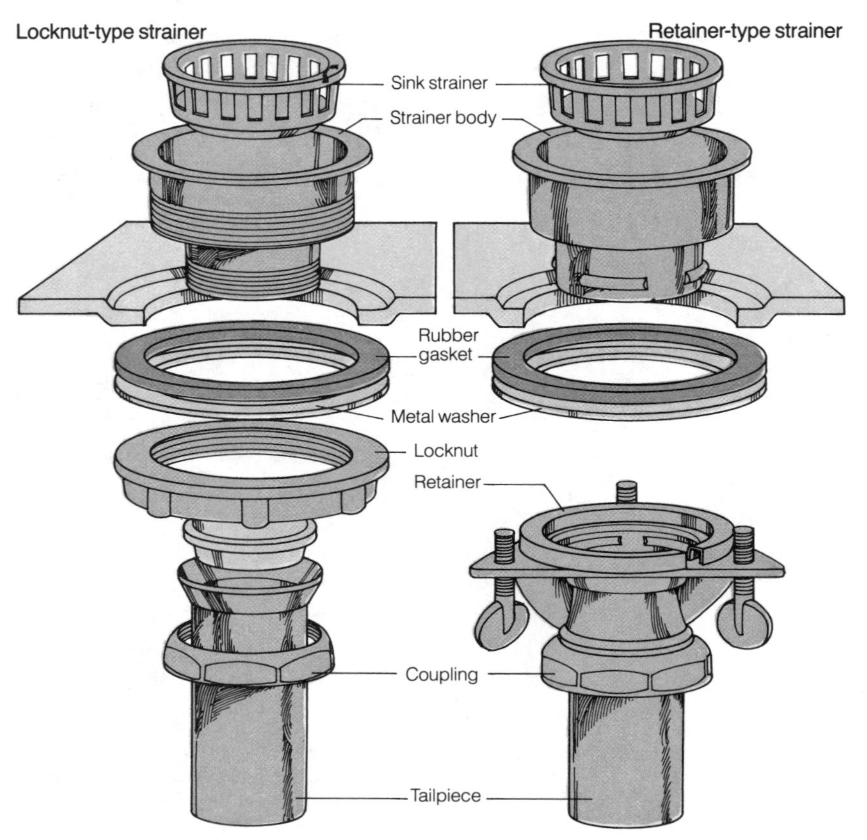

Drawing 28. Two types of sink strainers

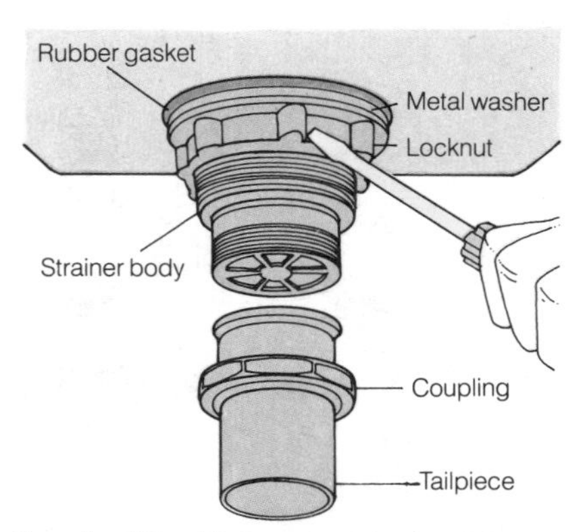

Drawing 29. To remove a locknut-type strainer, tap loose the lugs on the locknut, using a screwdriver and hammer.

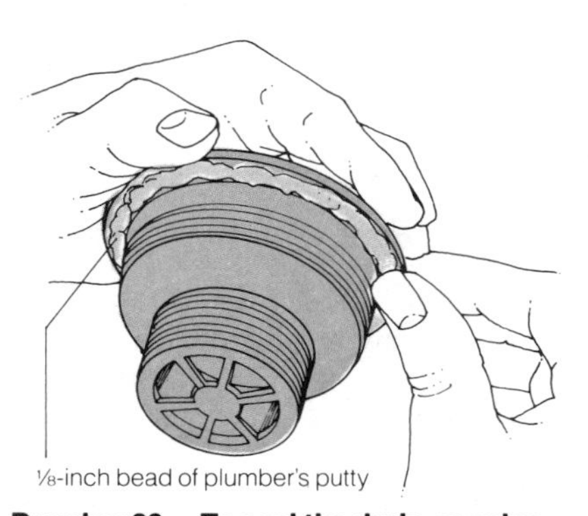

Drawing 30. To seal the drain opening, apply a bead of plumber's putty to the underlip of the strainer body.

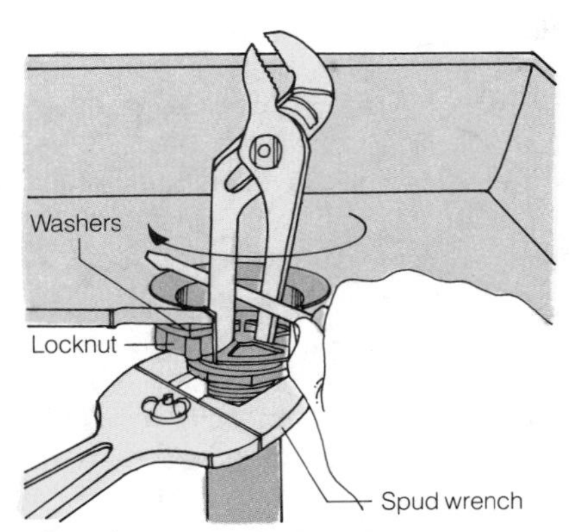

Drawing 31. To tighten a locknut strainer, use a screwdriver and pliers for counterforce from above.

Faulty Sink & Tub Pop-ups

What was once an old-style rubber stopper on a chain, plugging our sinks and tubs, is now likely to be a pop-up assembly (see **Drawing 32**). As its name implies, it pops up or down to open or close the drain, depending on the position of the lift rod. The lift rod works through the pivot rod to raise and lower the stopper. Though it sounds too simple ever to go wrong, the pop-up mechanism has several moving parts that need adjusting every so often. A faulty connection between the stopper and sink—the number one pop-up problem—can have several different causes.

Working on a sink pop-up. If the pop-up stopper isn't seating snugly, pull it out. Some stoppers (see **Drawing 33**) sit atop the pivot rod and simply lift out (A). Others (B) require a slight twist to free them because a body slot hooks them to the rod. Still others (C) are attached to the pivot rod.

Clean the stopper of any hair or debris. Check its rubber seal; if it's damaged, pry it off and slip on a new one. Also make sure the flange and the putty beneath it are in good shape. If the pop-up still doesn't seat correctly, loosen the clevis screw, push the stopper down, and retighten the screw in the next higher hole. When the drain is closed, the pivot rod should slope slightly uphill from clevis to drain body.

If the stopper is so tight that it doesn't open far enough for proper drainage, you'll need to reset the pivot rod by squeezing the spring clip and inserting the rod in the next lower hole.

If water drips from around the pivot ball, try tightening the retaining nut (see **Drawing 34**) that holds the ball in place. Still leaking? Replace the gasket or washer (or both) inside the assembly. Retighten the retaining pivot-rod nut and adjust the pivot rod so the pop-up seats properly.

Adjusting a tub pop-up. Remove the pop-up stopper (the options are the same as for a sink stopper). Then remove the tub's overflow plate and pull the entire assembly through the overflow (see **Drawing 35**). If the stopper doesn't seat properly, loosen the adjusting nuts and slide the middle link up the striker rod. The striker spring rests unattached atop the rocker arm.

For a sluggish drain, on the other hand, lower the link. Before reassembling, pull out the stopper, clean it, and check the flange for trouble spots.

NOTE: Some tubs have a strainer and internal plunger instead of a pop-up assembly. The repairs on the middle link are identical.

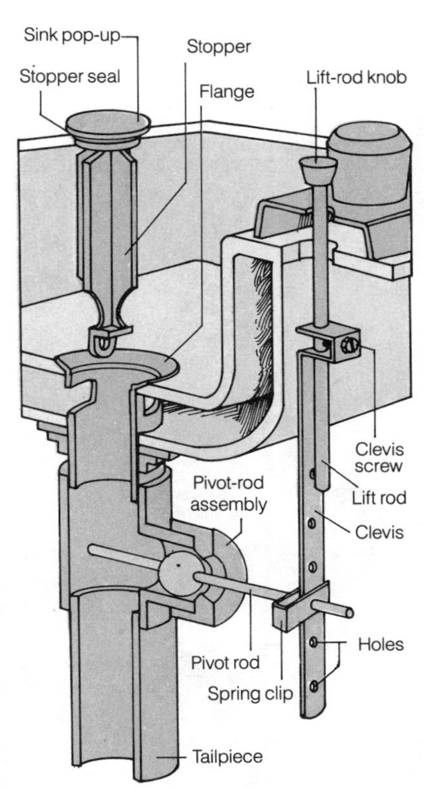

Drawing 32. Sink pop-up assembly opens and closes the drain when the lift rod is moved up or down.

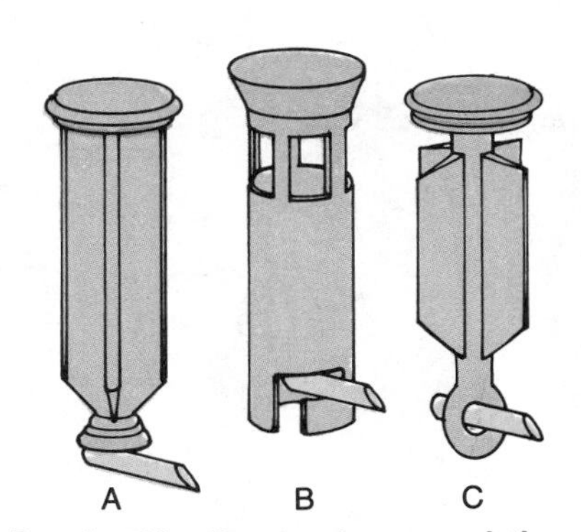

Drawing 33. Pop-up stopper varieties: unattached (A), slotted (B), and attached to the pivot rod (C).

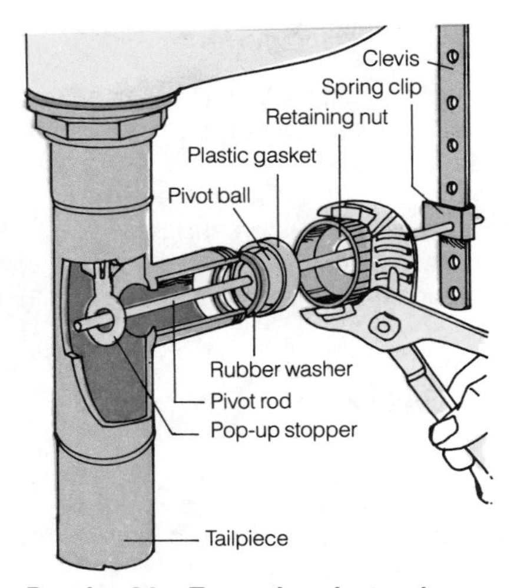

Drawing 34. To repair a pivot-rod assembly, tighten the retaining nut or replace the gasket or washer.

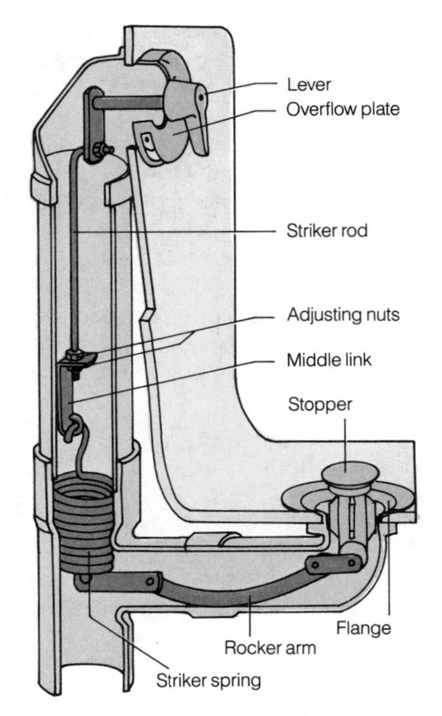

Drawing 35. To adjust a tub pop-up assembly, remove the overflow plate and pull the assembly out the overflow hole.

Valve Leaks

Valves, like faucets, control the flow of liquid through pipes by means of a basically simple mechanism: a handle that drives a stem down into its base to reduce or shut off flow. You'll find different types of valves for different uses: some restrict the flow even when fully open; others allow unrestricted flow when open. The three most commonly used types are gate, globe, and angle valves.

Valves differ also in the materials from which they're made. Those used in home plumbing are usually of cast bronze, though plastic valves are sold for use with plastic pipe. Brass valves are installed on gas lines, but can also be found on some water lines.

Gate valves

Used as the main shutoff valve in a residential system, a gate valve (see **Drawing 36**) completely shuts off or opens up a supply pipe; it's not designed to adjust flow. This valve has a tapered wedge at the end of a stem that moves up or down across the water flow.

Since it takes a half dozen or more turns to fully open a gate valve, many people tend to open the valve only partially, which can damage the valve mechanism. Slightly opening a gate valve permits partial flow, but the pressure of the water moving across the wedge wears down its surface, resulting in an imperfect seal and a leaking valve. For this reason, gate valves should never be operated partially opened. If used correctly, these valves can last for many years.

Globe valves

Unlike a gate valve, a globe valve (see **Drawing 36**) is designed to reduce water pressure: two half partitions change the direction of flow, slowing the water down.

Like a compression faucet (see pages 13–14), the globe valve has a stem that forces a disc into the valve seat. By turning the handle of a globe valve, you enlarge or diminish the opening for the water to pass through. This valve is the easiest kind to repair. Most often, the disc or seat washer is defective and needs replacement. It's a simple task; see "Repairing a leak around the stem," opposite.

Water supply pipes are usually equipped with globe valves, which can withstand frequent opening and closing under high pressure.

Angle valves

An angle valve (see **Drawing 36**) is similar to a globe valve, except that the water inlet and outlet are at a 90° angle to each other and there are no partitions. Water flow is less restricted than in a globe valve because the water makes only one turn instead of two. Since an angle valve eliminates the need for an elbow, these valves are often used where a pipe bends around a corner.

CAUTION: Before doing any work, shut off the water at the main shutoff valve (page 23). Open the nearest faucet to drain the pipes. Place a bucket under the valve to be worked on, to catch any remaining water.

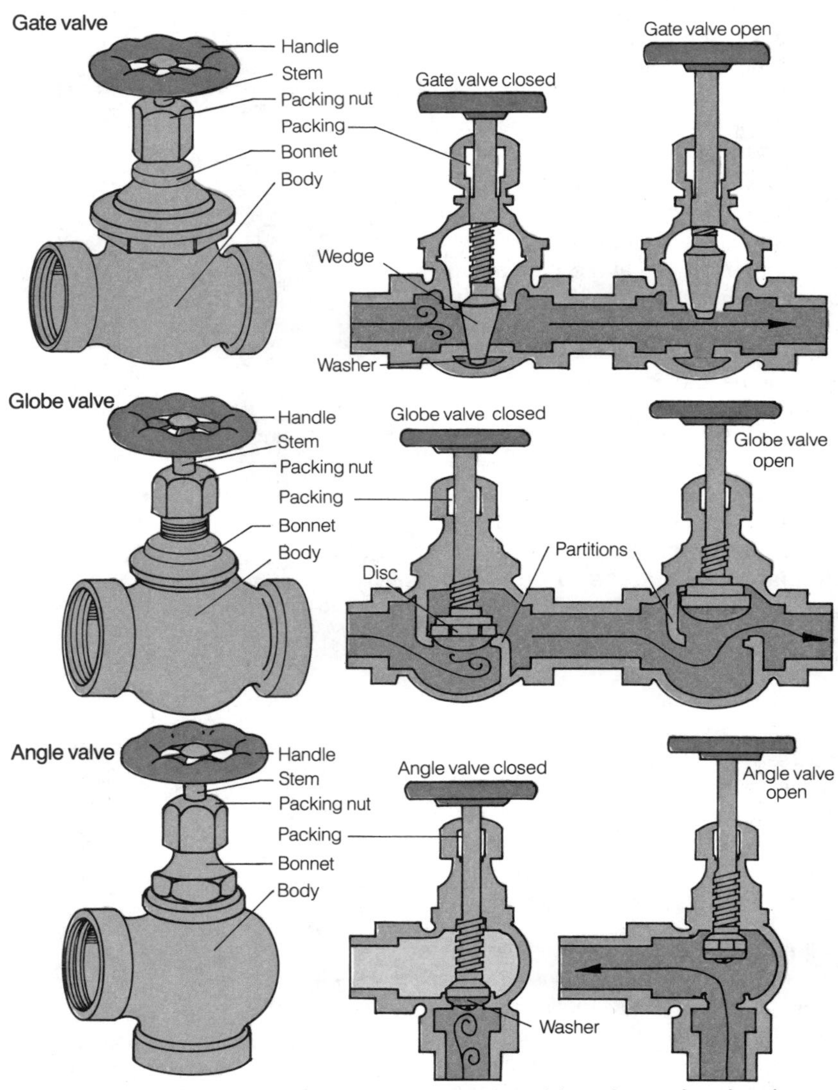

Drawing 36. Three common valve types: gate valve, globe valve, and angle valve.

Repairing a leak around the stem

The most common valve problem is leaking around the stem. The usual causes are compressed packing, a faulty seat washer or disc, an obstruction in the valve, or damage to the valve seat.

Despite the differing internal designs, the repairs for each valve type are basically the same.

To repair a valve, use an adjustable-end wrench to loosen and remove the packing nut beneath the valve handle. Examine the packing—if it's compressed, it can't do its job and you'll need to clean away all the old packing and wrap new graphite-impregnated twine around the base of the stem (see **Drawing 37**).

If this doesn't stop the leak, unscrew the valve stem and bonnet from the body. Inspect the seat washer or disc at the bottom of the stem. If it's faulty, unscrew the locknut, remove the washer or disc (see **Drawing 38**), and replace it with a new one—an exact duplicate.

If the new washer doesn't stop the leak, chances are there's an obstruction in the valve body. If that's the case, you'll need to clean the obstruction away. To do this, remove the valve stem as described before, and use a toothpick to clear away any blockage.

You may also need to detach the valve from its connecting pipe and clean the inside of the valve body with a stiff brush and soapy water.

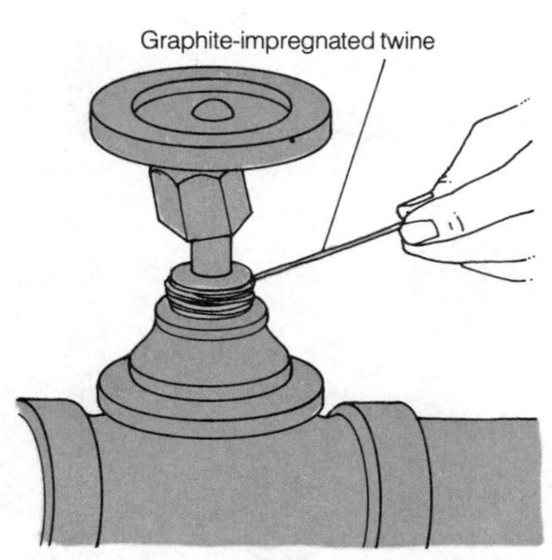

Drawing 37. To replace valve packing, remove the packing nut and clean away the old packing, then wind on new packing.

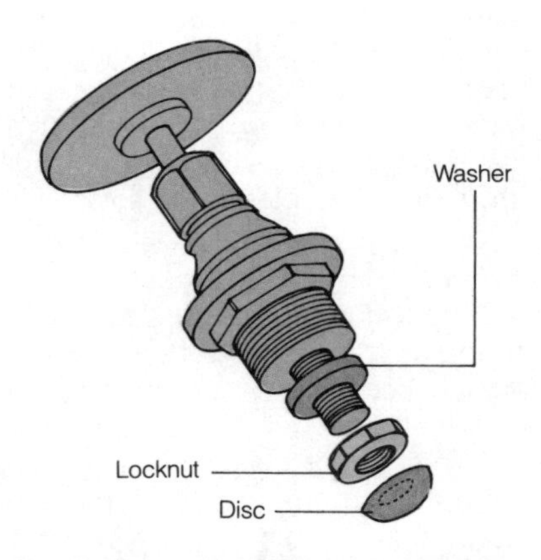

Drawing 38. To replace the seat washer or disc, remove the valve stem from the body and take off the locknut.

Turning off the water

You'll need to turn off the water whenever you want to work on your house's plumbing, whether it's a routine repair or an emergency.

Fixture shutoff valve. First look for shutoff valves for the fixture or appliance you need to work on (see **Drawing A**). These valves are usually directly under the fixture at the point where the water supply connects to it. You might make a project of installing shutoff valves on any fixtures in your home that don't have them (see pages 70–71).

Main shutoff valve. If the plumbing problem is not with a particular fixture, or if the fixture doesn't have its own shutoff valves, use the main shutoff valve (see **Drawing B**) to turn off the water supply to the entire house. The main shutoff is usually a gate valve located between the water meter and the house foundation. In cold climates, look just inside the foundation wall in the basement or crawl space. It's a good idea to know exactly where the main shutoff valve is and to test it before any trouble arises.

If the main shutoff valve itself needs work, call your water company; a special tool is required to shut off the water before it reaches that valve.

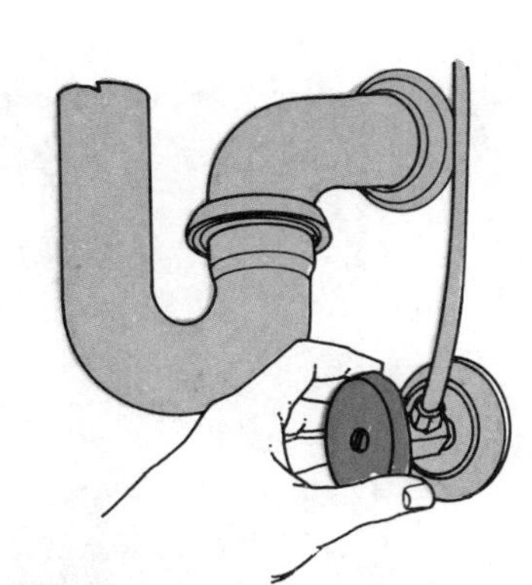

Drawing A. Sink shutoff valve

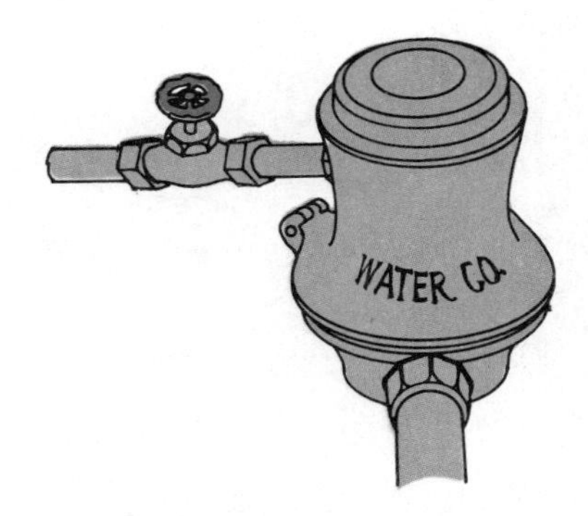

Drawing B. Main shutoff valve

Trap Problems

Traps are the workhorses of the drainage system (see page 8). Traps remain filled with just enough water to keep toxic sewage gases from coming up your drainpipe. Unfortunately, by the nature of their shape and function, traps are often the first plumbing part to cause clogs or leaks.

If the trap of a sink, tub, or shower leaks, the cause could be corrosion or stripped fittings. If there's a clog and you have attempted to clean it out (see pages 26–27), you can suspect a mineral buildup inside the trap. For any of these problems, the solution is to install a new trap and possibly a new tailpiece.

Traps come in assorted sizes and shapes (P traps in fixed or swivel types are most common—see **Drawing 39**) and are held together by a combination of slip-joint couplings. Although the S trap shown on page 8 may be found in older homes, this type is no longer permitted by codes in many parts of the country. (If you have S traps in your home, you're not required to replace them with P traps unless you're remodeling or altering part of your plumbing system.)

Ordinarily, home plumbing systems use P traps. A "fixed" P trap is a continuous length of pipe in the shape of a P. A "swivel" P trap is actually a J bend trap and an elbow that together form a P and are called, for the sake of convenience, a P trap rather than a "J bend and elbow."

Trap materials vary, as do sizes and shapes. Traps may be of brass, galvanized steel, or plastic, depending on local plumbing codes. Chrome-plated traps, the most expensive of the lot, last the longest.

CAUTION: Before doing any work, turn off the water at the fixture shutoff valves or the main shutoff valve (page 23). Open the faucet to drain the pipes.

Replacing a trap

New traps are sold as complete units with washers, threaded couplings, and the fitting itself.

The first step in removing the old trap is to place a pail beneath the trap and remove the cleanout plug, if there is one.

Use a tape-wrapped wrench to loosen the couplings that attach a fixed P trap to the tailpiece and drainpipe (see **Drawing 40**) and a swivel P

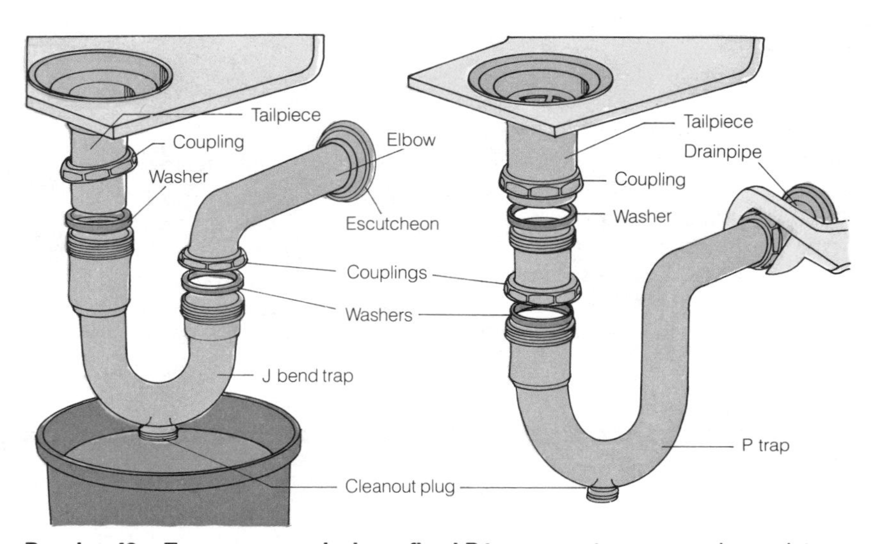

Drawing 40. To remove a swivel or a fixed P trap, use a tape-wrapped wrench to loosen the connecting couplings at the tailpiece and elbow or drainpipe. Carefully pull the trap away.

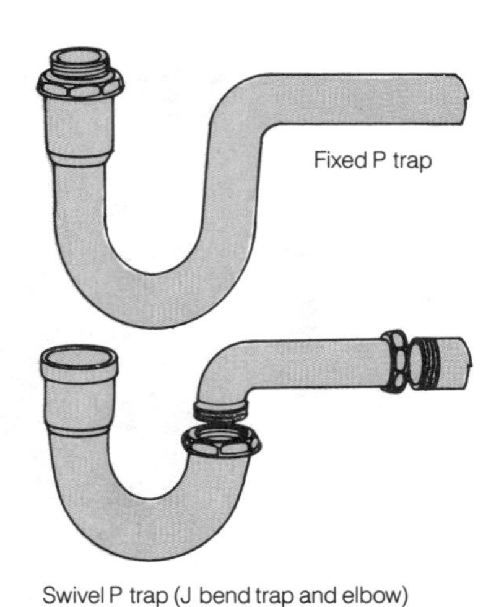

Drawing 39. Types of traps

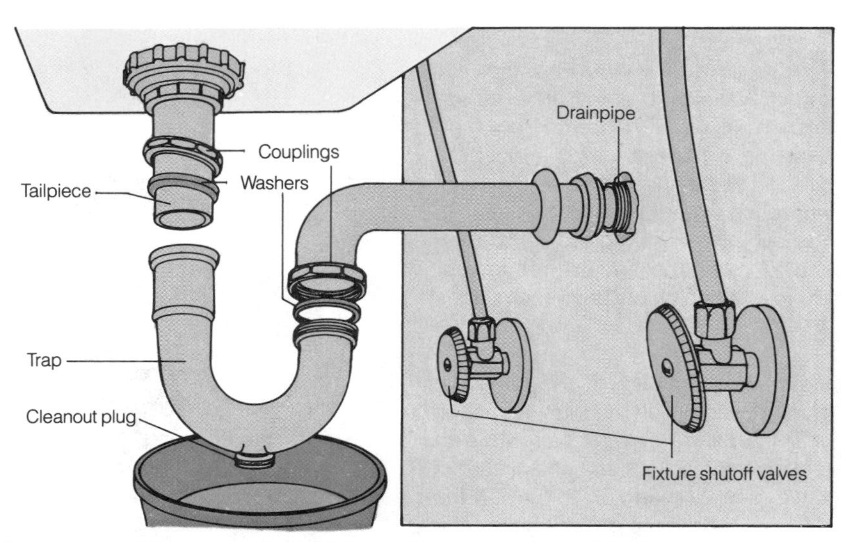

Drawing 41. To attach a swivel P trap, slide the couplings and washers—in that order—into place and carefully tighten with a tape-wrapped wrench.

(J bend) to the elbow as well. Pull off the trap.

To attach a new swivel P trap, slide the new couplings and washers—in the order shown (see **Drawing 41**)—over the sink tailpiece, connecting drainpipe, and elbow. Coat the threads of the drainpipe with plumber's grease; attach the elbow. Set the trap in place and tighten the couplings at both ends by hand. With the tape-wrapped wrench, finish tightening, being careful not to strip or overtighten the couplings. Turn the water back on and check all connections for leaks.

Replacing a tailpiece

A tailpiece that is cracked or corroded must be replaced, and fortunately replacements are available separately, not just as part of a faucet assembly.

To remove the tailpiece, unscrew the couplings that fasten it to the trap and the sink drain, and push the tailpiece down into the trap. Loosen the couplings at the drainpipe or elbow and turn the entire trap, at the drainpipe, a quarter turn—just far enough to allow room to remove the tailpiece. You can now lift the old tailpiece out of the trap (see **Drawing 42**) and replace it with a new one.

Coat the threads of the new tailpiece with plumber's grease to ensure a watertight seal. Hand tighten all couplings, then use a tape-wrapped wrench to snug them up. Restore water pressure and look for leaks.

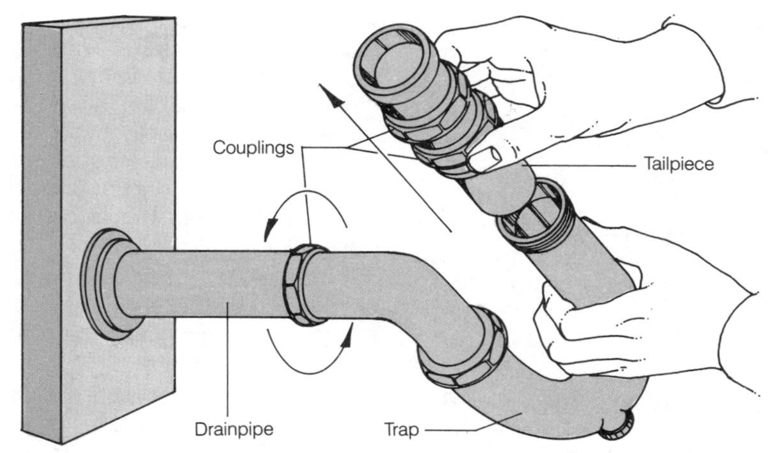

Drawing 42. To remove a tailpiece, loosen the couplings, push the tailpiece down into the trap, and turn the trap counterclockwise a quarter turn.

Preventing clogged drains

No plumbing problem is more common or more frustrating than a clogged drain. Kitchen sink drains clog most often because of a buildup of grease, and food particles that get caught in the grease. Hair and soap are usually at fault in bathroom drains. Drains can usually be cleared easily and inexpensively (see pages 24–31), but taking some simple precautions will help you avoid the inconvenience of stop-ups.

Proper disposal of kitchen waste will keep sink drain clogs to a minimum. Be especially careful not to pour grease down the kitchen sink. Another villain is coffee grounds—throw them out, don't wash them down.

Be sparing with chemical cleaners, particularly if you have brass, steel, or cast iron traps and drainpipes; some caustic chemicals can corrode metal pipes. (Plastic drainpipes are more resistant to damage from caustic chemicals.) If used no more than once every few months, cleaners containing sodium hydroxide or sodium nitrate can be safe and effective.

When you use a drain cleaner, be sure to follow the instructions on the package carefully. You'll need to let the cleaner sit in the bend of the trap awhile for it to be effective. Be careful not to splash it about or get any on your skin. Rinse the area thoroughly after using the cleaner so no traces of it remain.

Clean floor drain strainers. Some tubs, showers, and basement floor drains have strainers held in place in the drain outlet by screws. You can easily remove these strainers and reach down into the drain to clear out accumulated debris. And be sure to scrub the strainer itself.

Clean pop-up stoppers in the bathroom sink and tub regularly. Some sink pop-ups lift straight up and out; others require a half-turn to the left; still others must be disengaged from below. (For details on removing pop-up stoppers, see page 21.) For most types of sink pop-ups, it's a simple matter to lift them out once a week to remove collected hair and rinse them off.

Every few months, unscrew the overflow plate on a tub and pull up the pop-up assembly (see page 21) to reach the spring or rocker arm where hair accumulates. Remove collected hair and rinse thoroughly.

Finally, flush the drain-waste and vent (DWV) system whenever you go up onto the roof of your house to clean out drains or gutters with a garden hose. Run the hose down into all vents and give them a minute or two of full flow.

Clogged Sinks

A stopped drain isn't just an inconvenience; it can sometimes be an emergency.

It's a good idea to take action against clogs before the situation becomes dire (for information on preventing clogs, see page 25). At least be alert to the warning sign of a sluggish drain—it's easier to open a drain that's slowing down than one that's stopped completely.

When it's too late for preventive medicine, a dose of scalding water—especially effective against grease buildups—may be treatment enough. If not, it could be that something foreign—a hairpin, button, coin, or small utensil—has slipped down the drain. To check, remove (and thoroughly clean) the sink pop-up stopper (see page 21) or the strainer (see page 20) in the drain.

Usually, a clog will be close to the fixture. You can determine this by checking the other drains in your home. If more than one won't clear, something is stuck in the main drain (pages 30–31). Otherwise, you're probably dealing with a clog in the trap or drainpipe. Use one or more of the tried-and-true weapons in the drain-cleaning arsenal: the plunger, a chemical cleaner, or the snake.

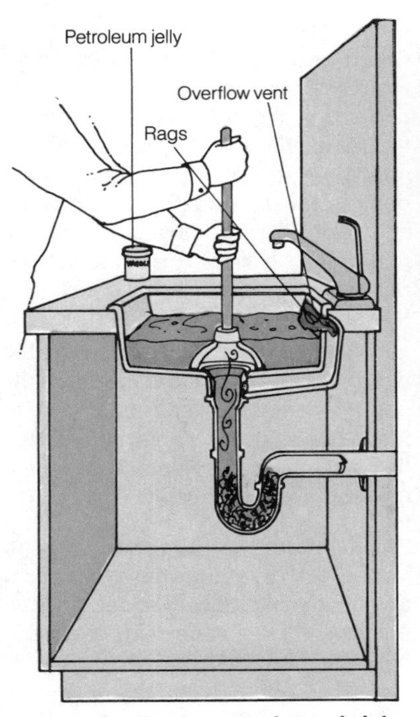

Drawing 43. To plunge a clogged sink, pump the petroleum-jelly-coated plunger vigorously up and down, 15 to 20 times.

The plunger

What's the first reaction to a clog? Reach for the plumber's helper—the plunger. The plunger is a good drain-cleaning tool, but it often fails to work because it's incorrectly used. Don't make the typical mistake of pumping up and down two or three times, expecting the water to whoosh down the drain.

Though no great expertise is needed to use this simple tool, here are a few tips to guide you toward efficient plunging:

- *Choose a plunger* with a suction cup large enough to cover the drain opening completely.
- *Fill the clogged fixture* with enough water to cover several inches of the plunger cup.
- *Coat the rim* of the plunger cup with petroleum jelly to ensure a tight seal.
- *Block off all other outlets* (the overflow vent, the second drain in a double sink, adjacent fixtures) between the drain and the clog with a wet cloth.
- *Insert the plunger* into the water at an angle, so that no air remains trapped under it.
- *Use 15 to 20 forceful strokes,* holding the plunger upright (see **Drawing 43**).
- *Repeat the plunging* two or three times before giving up.

Drawing 44. When using a chemical drain cleaner, wear gloves, keep your face away from the drain, and ventilate well.

Chemical drain cleaners

If the water is draining somewhat but plunging has failed to open the drain completely, you may want to try one of the many drain cleaners on the market. Many people, in fact, rely on chemicals as a preventive measure, routinely using them before a clog forms. Such measures will probably prevent a clog, but they're not advisable because the chemicals may eventually damage your pipes and inhibit beneficial bacterial growth in a septic tank.

Whenever you use chemicals, do so with caution and in a well-ventilated room (see **Drawing 44**).

- *Don't use a chemical cleaner* if the blockage is total, especially if the fixture is filled with water. It won't clear the blockage, and you'll face another problem—getting rid of the caustic water.
- *Wear rubber gloves* to prevent the chemical from burning your skin.
- *Read labels,* and match cleaners with clogs. Alkalis cut grease; acids dissolve soap and hair.
- *Don't mix chemicals.* Mixing an acid and an alkali cleaner can cause an explosion.
- *Don't look down the drain* after pouring in a chemical. The solution frequently boils up and gives off toxic fumes.
- *Never plunge* if you've recently used a chemical.

The plumber's snake

Should the plunger and chemical treatment fail, look to the snake (also called a drain-and-trap auger). This tool (shown on page 11) is a very flexible metal coil that you feed through the pipes until it reaches the clog.

Check for the clog close to the drain first, by inserting the snake through the drain. If that doesn't clear the pipes, put the snake through the trap cleanout, if there is one. And if that doesn't work, remove the trap entirely so the snake can reach through the drainpipe to clear a clog that's farther from the drain. Once you remove the trap, though, you may find that you don't even need the plumber's snake to dig deeper. Just clean out the trap with a brush and soapy water and your troubles may be over.

Through the drain. Remove the pop-up stopper (page 21) or sink strainer (page 20). Insert the snake in the drain opening (see **Drawing 45**) and twist it down through the trap until you reach the clog.

Through the cleanout. If the trap has a cleanout, place a pail beneath the trap to catch the water, and remove the nutlike cleanout plug from the bottom of the trap. Insert the snake (or a bent hanger) through the cleanout (see **Drawing 46**) to work at the clog. Direct the snake back up toward the drain, or angle it toward the wall to reach a deeper blockage, turning it as you go.

Through the drainpipe. Remove the trap as outlined on pages 24–25, unfastening the couplings with a tape-wrapped wrench. Pull the trap downward and spill its contents into a pail. Insert the snake into the drainpipe at the wall. Feed it as far as it will go (see **Drawing 47**) until it hits the clog. Clean out the trap.

If the snake doesn't succeed, the clog is probably too deep in the pipes to reach through the drainpipe. This means you're dealing with a main drain clog (see pages 30–31) that needs to be attacked through the main soil stack, main cleanout, or house trap.

How to use a snake. Feed the snake into the drain, trap, or pipe until it stops. If the snake has a movable hand grip, position it about 6 inches above the opening and tighten the thumbscrew. Rotate the handle to break the blockage. As the cable works its way farther into the pipe, loosen the thumbscrew, slide the hand grip back, push more cable into the pipe, tighten the thumbscrew again, and repeat. If the snake doesn't have a hand grip, maneuver the cable by simultaneously pushing and twisting until it hits the clog.

The first time the snake stops, it probably has hit a turn in the piping rather than the clog. Guiding the snake past a sharp turn takes patience and effort. Keep pushing it forward, turning it as you do. Once the head of the snake hooks some of the blockage, pull the snake back a short distance to free some material from the clog, then push the rest on through.

After breaking up the clog, pull the snake out slowly and have a pail ready to catch any gunk the snake brings with it.

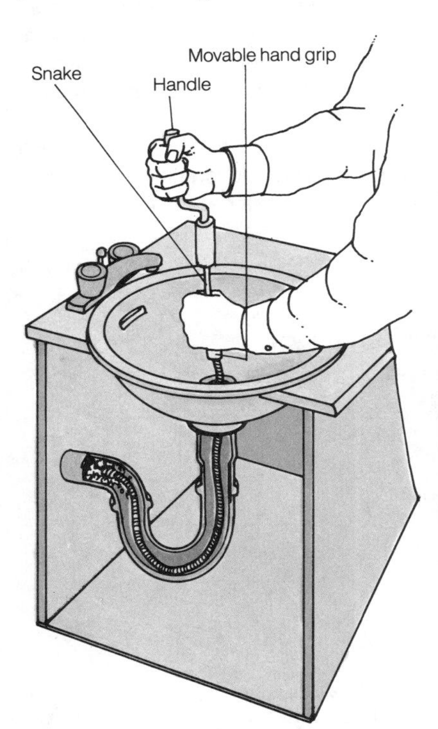

Drawing 45. Snaking through the drain can dislodge a clog or small item that is blocking the flow.

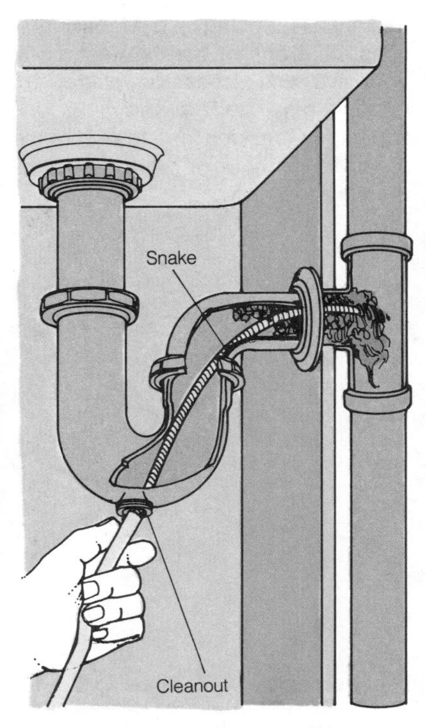

Drawing 46. Snaking through the cleanout—if the trap has one—can get at a clog from below.

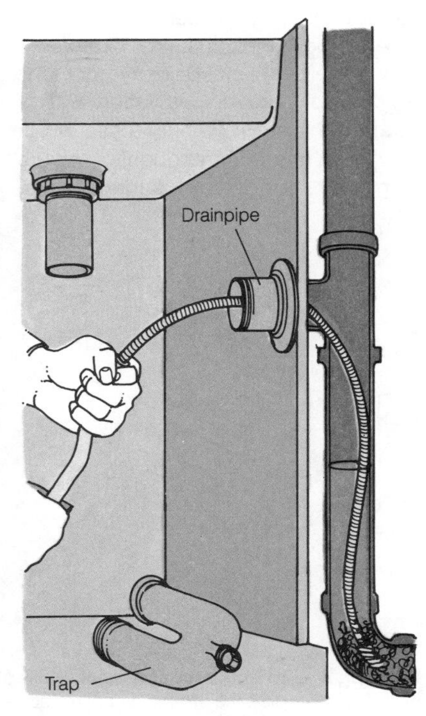

Drawing 47. Snaking through the drainpipe means removing the trap to reach a deep-down blockage.

Face lifts for fixtures

As time takes its toll, the beauty of your plumbing fixtures will fade. Once-gleaming porcelain and fiberglass surfaces tend to dull, discolor, or become damaged. Separations can appear where a bathtub joins the wall.

But before you put a stained, chipped, or cracked fixture out to pasture, consider giving it a face lift with some bleach, touch-up paint, glue, or caulk. Often such cosmetic repairs can extend the life of an aging fixture, postponing expensive replacements.

Stains and mineral deposits. You can renew a fixture by removing discolorations, rust stains, and mineral deposits. An effective treatment for all of these problems is a liquid chlorine bleach solution. Use a mixture of cream of tartar and peroxide to improve the appearance of an enameled fixture.

To remove rust stains from porcelain or fiberglass, try rubbing the stain with a cut lemon, or applying lemon juice. If the fixture is seriously stained, use a 5 percent solution of oxalic acid or a 10 percent solution of hydrochloric acid. Apply the acid solution with a cloth and leave it on only a second or two; then rinse off thoroughly. Be sure to protect your skin by wearing gloves.

Small scratches and chips. An aerosol can slips from your hand and suddenly you have a chipped sink. Don't despair. You can cover a small scratched or chipped area of porcelain or fiberglass by building up thin coats of enamel or epoxy paint (see **Drawing A–1**), available in touch-up kits in many matching colors.

Before applying the paint, clean the surface of the chipped area with alcohol, making sure it's clean, dry, and dustfree. Using a small brush, apply several coats of touch-up paint, blending it toward the edges of the chip. Allow the paint to dry for one hour between coats. Don't expect perfection—you may be able to see where the touch-up was done, once it dries.

Large chips. If the corner or edge of a porcelain or fiberglass fixture has broken off, and if you have a chip that fits in the place it came from, you can glue it back in place with epoxy resin. Again, be sure the surfaces are entirely clean and dry. Depending on the label directions, coat one or both surfaces with adhesive and press the pieces together firmly (see **Drawing A–2**). Use a clamp or masking tape to secure the repair for an hour or more. Keep the area dry overnight.

Separated bathtub-wall joint. One of the most common bathroom repairs is sealing cracks in the joint between the bathtub and the wall. It's a chore you'll have to repeat every year or so because the weight of the tub changes as it's filled and emptied again and again.

The simplest way to seal the troublesome joint is with flexible waterproof caulking compound, commonly called plastic tub and tile sealer. The sealer comes in a tube.

Before applying sealer, scrape away the old caulking. Clean and dry the area thoroughly to ensure a good seal. Holding the tube at a 45° angle, slowly squeeze the sealer into the tub joint (see **Drawing B–1**), using a steady, continuous motion. If you can do each side of the tub without stopping, the line of caulking will be smoother and neater. Wait at least 24 hours before using the bathtub.

If you find caulking won't last in the bathtub-wall joint, apply quarter-round ceramic edging tiles (see **Drawing B–2**). Available in kits, the tiles are easy to install around the rim of the tub; use the caulking compound just described as an adhesive. Be sure to scrape away old caulking and clean and dry the area before you begin.

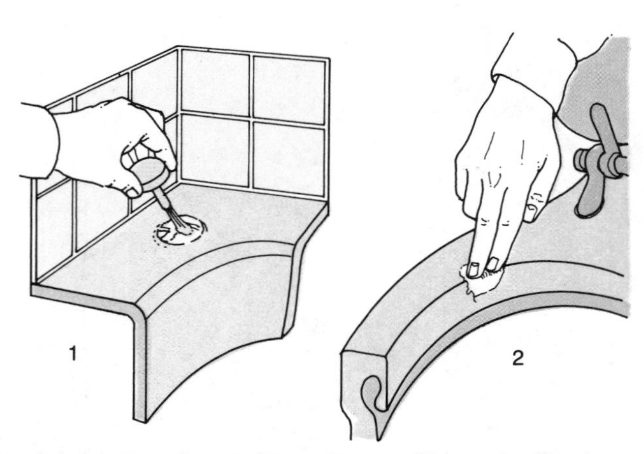

Drawing A. To make repairs, apply several thin coats of touch-up paint (1), or glue the broken piece in place with epoxy resin (2).

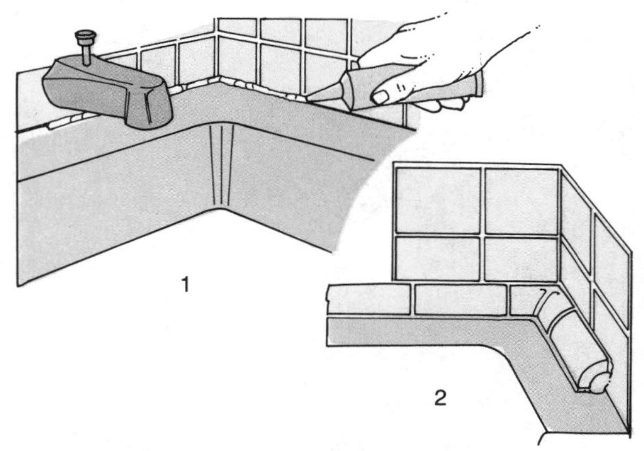

Drawing B. To cure a tub-wall separation, run a continuous bead of waterproof caulking (1) or install quarter-round ceramic edging tiles (2) along the joint.

Plugged Tubs & Showers

Like sink drains, bathtub and shower drains can eventually plug up. When they do, first see whether other fixtures are affected. If they are, work on the main drain (see pages 30–31). If only the tub or shower is plugged, work on it.

Tub with a P trap

Remove the overflow plate and pull the pop-up or plunger assembly (see page 21) out through the opening. Feed the snake down through the overflow pipe and into the P trap (see **Drawing 48**). Use the twisting, probing method described on page 27. This should clear the drain. If not, remove the trap or its cleanout plug and insert the snake toward the main drain.

Tub with a drum trap

Instead of a common P trap, bathtubs in older structures may have a drum trap (see **Drawing 49**) located alongside the tub. If there's a clog, bail all standing water from the tub so that water from the trap doesn't overflow onto the floor. Slowly unscrew the drum trap cover with an adjustable wrench. Watch for water welling up around the threads; have absorbent rags ready. Remove the cover and rubber gasket on the trap and clean any debris from the trap itself. If you find no obstruction there, work the snake through the lower pipe toward the tub. Still no clog? Direct the snake in the opposite direction (away from the tub) toward the main drain.

Shower drain

Unscrew the strainer over the drain opening. Probe the snake down the drain and through the trap until it hits the clog (see **Drawing 50**).

For deep clogs in shower drains, a garden hose (see **Drawing 51**) is often more effective than a snake. Use a threaded adapter to attach the hose to a faucet. Push the hose deep into the drain trap and pack rags into the opening around the hose. While holding the hose and rags in the drain, turn the water to the hose alternately on full force and abruptly off.

CAUTION: Never leave a hose in any drain; a sudden drop in water pressure could siphon sewage back into the fresh-water supply.

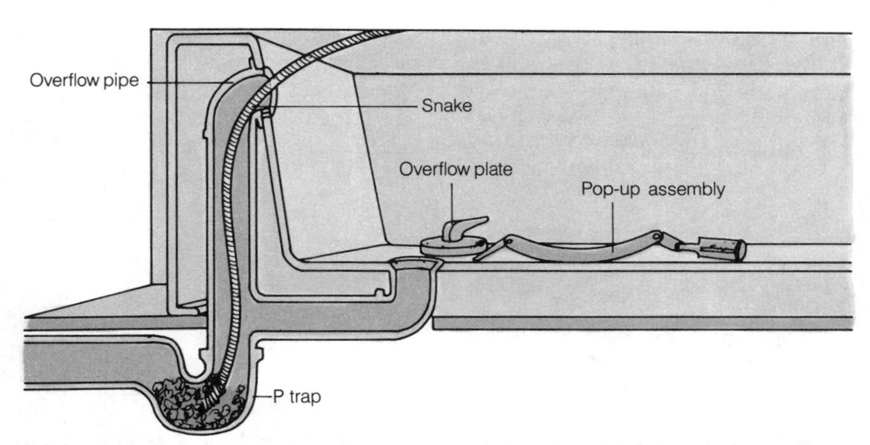

Drawing 48. To unclog a tub with a P trap, feed the snake down through the overflow pipe and maneuver it until it reaches the blockage.

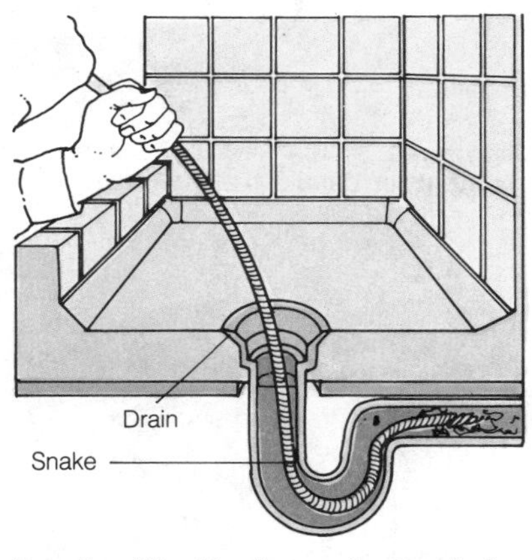

Drawing 50. To clear a shower drain, direct the snake down through the drain opening into the trap.

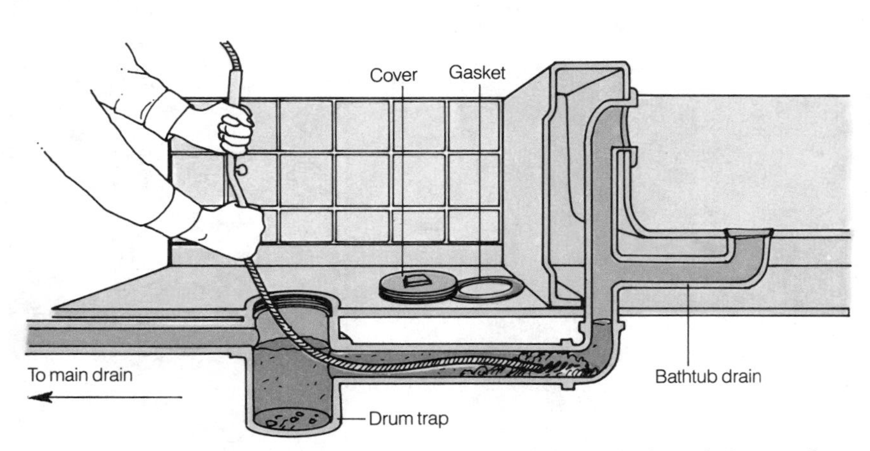

Drawing 49. To unclog a tub with a drum trap, work the snake through the trap, first toward the tub, then toward the main drain.

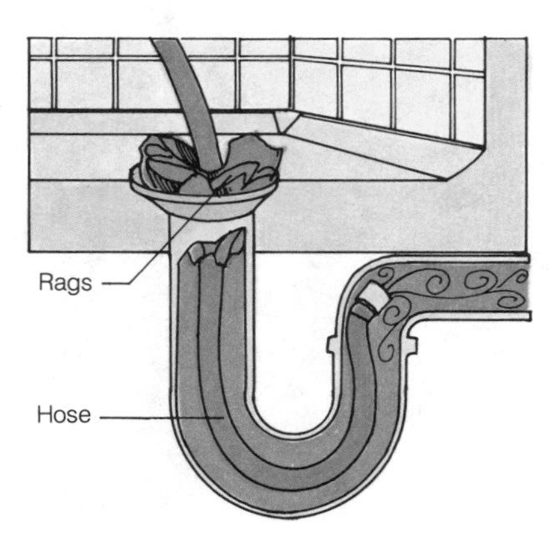

Drawing 51. Use a garden hose to open a shower drain if a snake doesn't get at the blockage.

Main Drain Clogs

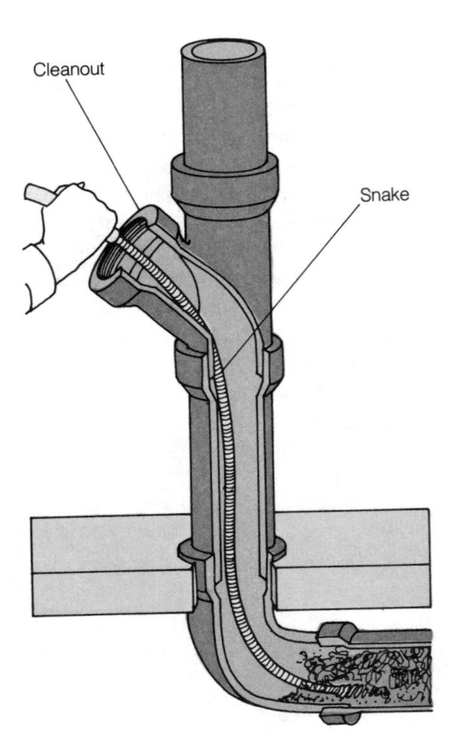

Drawing 52. A snake run through a cleanout can bend and maneuver.

If a clog is too deep in the pipes to get at from a fixture (as detailed on pages 26–29), the trouble lies somewhere in the drainage system that channels waste into the sewer. When you have a blocked soil stack or main drain, all the fixtures above the clog stop up. If there's a blockage in the vent stack (the pipe that keeps air flowing out the roof), wastes drain slowly, and odors from the pipes become noticeable in the house. (See page 8 for a drawing of these systems.)

To troubleshoot the clog, trace the pipes from the plugged fixtures to the main soil stack—the vertical pipe to which all the branches connect. You can clean out the soil stack from above or below. Cleaning it from above means feeding a long snake down the vent stack from the roof; take safety precautions, especially on a steep or slippery roof. Cleaning the soil stack from below—running a hose or snake through the main cleanout, or a snake through the house trap—means working with raw sewage. It's a very messy job; be forearmed with pails, mops, and rags.

If none of the methods described in this section works, you'll have to call a plumber or professional drain-cleaning firm. Their workers use heavy-duty, power-operated augers that reach and cut through blockages, usually quickly and efficiently.

Snakes, hoses, and balloon bags. Snakes—very long hand-operated ones and power ones—are the primary tools used in clearing main drain clogs. A snake (see **Drawing 52**) can probe and maneuver around bends to seek out a stubborn clog. You can use some type of snake whether you're approaching the clog through the soil stack from above, through a cleanout, or through the house trap.

The length of the snake needed depends on the height of the soil stack; snakes from 50 to 75 feet long are commonly used. The diameter of the snake varies according to the stack diameter, from a ½-inch snake in

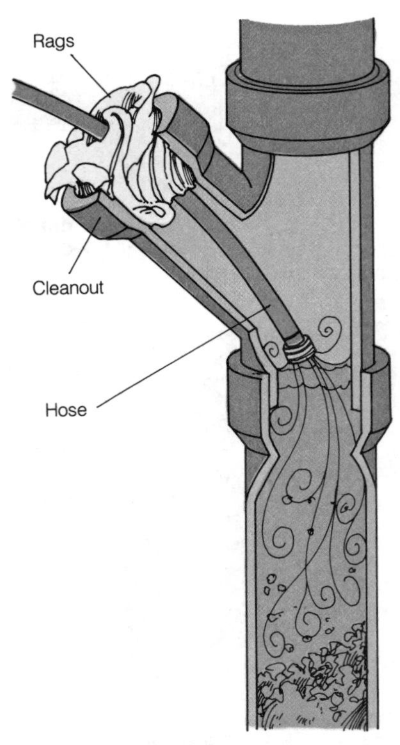

Drawing 53. A garden hose with rags can be more effective than a snake to force out a deep blockage.

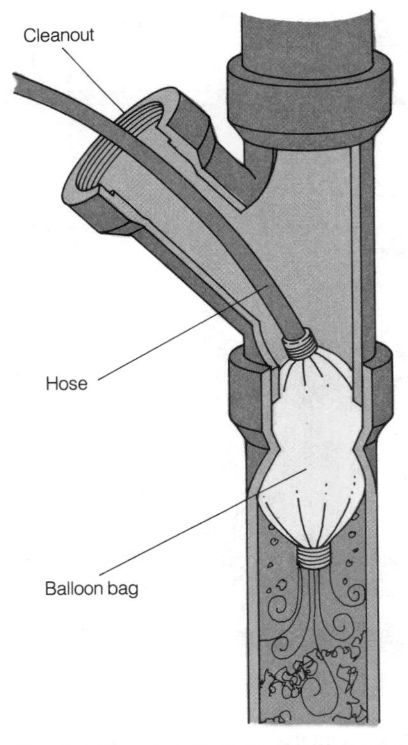

Drawing 54. A balloon bag attached to a hose nozzle delivers a powerful surge of pressure to clear out a clog.

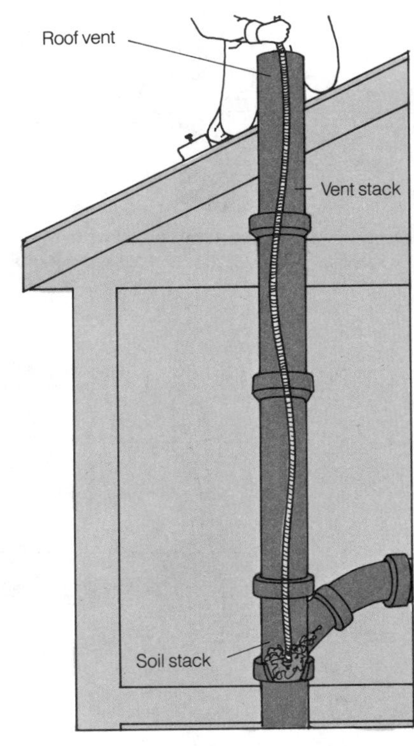

Drawing 55. Snaking through the vent stack to the soil stack means attacking a blockage from above with a long snake.

a 2-inch-diameter stack to a ¾-inch snake in a 4-inch-diameter stack.

If a hand-operated snake doesn't do the job, rent a power one (rental is often for a 3-hour minimum). Know your stack's diameter when you go to rent a snake. Use caution and work with a friend; a power snake can be hard to handle and dangerous to fingers and toes.

If you're trying to clear the clog by working through a cleanout, you have two additional options: a garden hose with rags stuffed around it (see **Drawing 53**), or a balloon bag attached to a hose nozzle (see **Drawing 54**). For more information on using a hose, see page 29.

Clearing the soil stack from above. Position yourself securely on the roof. Run an extra-long snake (length depends on the height of the stack) through the roof vent and down the vent stack (see **Drawing 55**), working it from side to side until it can go no farther. If the clog's not in the main soil stack, it's in the main drain to the sewer; try reaching it from the main cleanout or house trap.

Finding the main cleanout. Usually a Y-shaped fitting, the main cleanout (see **Drawing 56**), is near the bottom of the soil stack where the main drain leaves the house. Look for it in the basement or crawl space, or on an outside wall near a plumbing fixture.

In most newer buildings, there are several cleanouts—usually one wherever a branch of the drainage system makes a sharp turn. (The clog may be accessible from one of these branch cleanouts; try reaching the clog through one of them before opening the main cleanout.)

Opening the main cleanout. Set a large empty pail and spread some newspapers underneath the main cleanout to catch the waste water in the drainpipe. Use a pipe wrench to remove the plug (see **Drawing 57**)—open it slowly to control the flow of waste. Use a snake, hose, or hose with balloon bag to remove the obstruction; then flush with water. Coat the plug with pipe joint compound and recap the cleanout. If none of these methods is successful, move downstream to the house trap.

Working on the house trap. You can identify this fitting by its two adjacent cleanout plugs near where the main drain leaves the house. Generally the house trap is at ground level if the main drain runs under the floor.

Before opening the house trap, spread rags and newspapers around it to absorb any overflow. With a pipe wrench, slowly loosen the plug nearest the outside sewer line. Probe the trap and its connecting pipes with a snake (see **Drawing 58**). Be prepared to withdraw the snake and cap the trap quickly when water starts to flow. When the flow subsides, open both ends of the trap and clean it out with a wire brush. Recap and flush the pipes with water from an upstream cleanout.

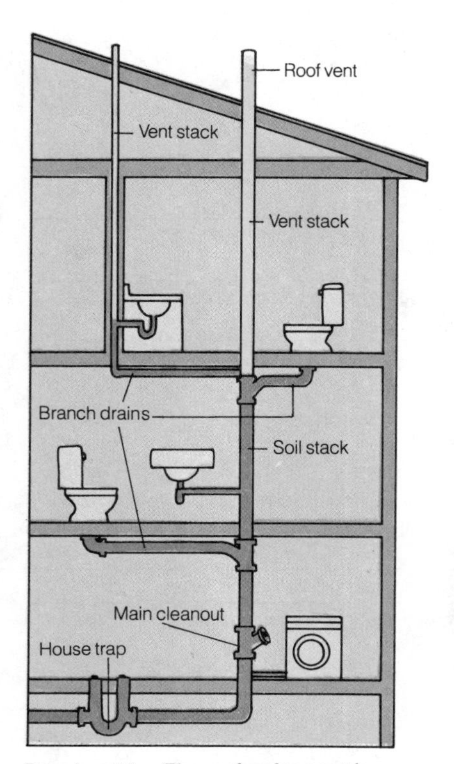

Drawing 56. The main cleanout is usually located at the bottom of the soil stack where waste enters the main drain.

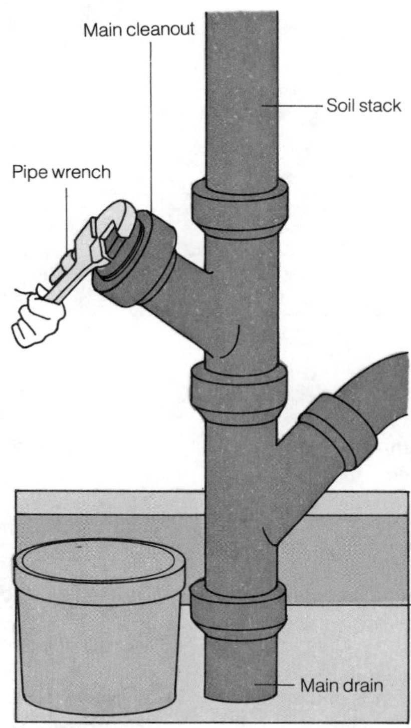

Drawing 57. When opening the main cleanout, be prepared with a large pail to catch the flood that may occur.

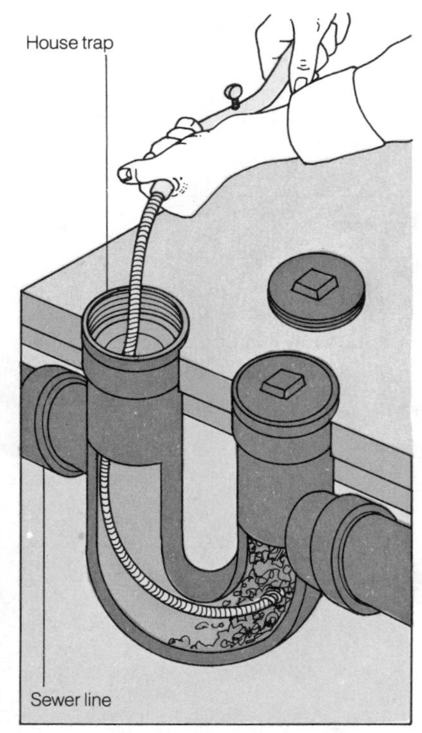

Drawing 58. When snaking through a house trap, work slowly, allowing water to drain gradually.

Guide to Toilet Repairs

The workings of a toilet remain a mystery to most people until something goes awry. Fortunately, what appears to be complex is, in fact, quite simple. Basically, there are two assemblies concealed under the lid: a ball cock assembly, which regulates the filling of the tank, and a flush valve assembly, which controls the flow of water from the tank to the bowl.

Here's the chain of events that occurs when someone presses the flush handle: The trip lever raises the lift wires (or chain) connected to the stopper. As the tank stopper goes up, water rushes through the valve seat into the bowl via the flush passages. The water in the bowl yields to gravity and is siphoned out the trap to the drainpipe.

Once the tank empties, the stopper drops into the flush valve seat. The float ball trips the ball cock assembly to let a new supply of water into the tank through the fill tube. As the tank water level rises, the float ball rises until it gets high enough to shut off the flow of water, completing the process. If the water fails to shut off, the overflow tube carries water down into the bowl to prevent the tank from overflowing.

Before you can tackle a toilet problem, you'll need to learn a little about the fixture's innermost components (see **Drawing 59**). Once you've done that, the chart should help you pinpoint a toilet troublemaker. On the following pages is information on fixing toilets that make noises, run, clog, and generally act up.

CAUTION: When you remove the tank lid, place it out of the way where it won't be hit by a falling wrench. Secondly, don't force a stubborn nut; oil it. This reduces the risk of slipping with the wrench and cracking the tank or bowl—a costly replacement.

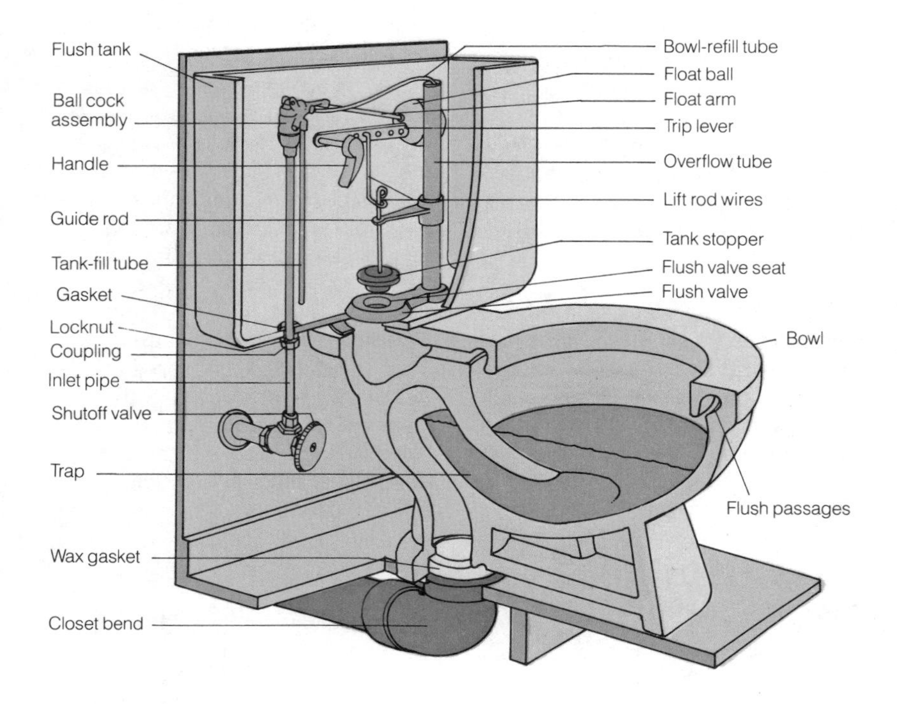

Problem	Causes	Remedies	Page
Noisy flush	Defective ball cock	Oil lever, replace faulty washers, or install new ball cock assembly	33
Continuously running water	Float arm isn't rising high enough	Bend float arm down or away from tank wall	34
	Water-filled float ball	Replace ball	34
	Tank stopper isn't seating properly	Adjust stopper guide rod or chain; replace defective stopper	34
	Corroded flush valve seat	Scour valve seat or replace it	34, 35
	Cracked overflow tube	Replace tube or install new flush valve assembly	35
	Ball cock valve doesn't shut off	Oil lever, replace faulty washers, or install new ball cock assembly	33
Clogged toilet	Blockage in drain	Remove blockage with plunger or closet auger	36
Inadequate flush	Faulty linkage between handle and trip lever	Tighten setscrew on handle linkage, or replace handle	37
	Tank stopper closes before tank empties	Adjust stopper guide rod or chain	34
	Leak between tank and bowl	Tighten locknuts under tank, or replace gasket	37
	Clogged flush passages	Poke obstructions from passages with wire	37
Sweating tank	Condensation	Install tank insulation or a tempering valve	37

Noisy Toilets

If your toilet is crying out for help with a high whine or whistle, the ball cock assembly may be to blame (see **Drawing 60**). The *diaphragm-type* ball cock assembly is the most conventional mechanism, using about 5 gallons a flush. The *two-way flush valve* has a handle that moves up for a light flush, down for a full flush. Its two stoppers regulate half or full-tank flow. The *float-cup ball cock* eliminates the need for a float arm and ball; a simple device, it helps prevent silent leaks.

Be sure any replacement assembly you buy has been tested and approved to prevent backflow.

CAUTION: Before doing any work, turn off the water at the fixture shutoff valve or the main shutoff valve (see page 23). Flush the toilet twice to empty it, and sponge out any water.

Replacing ball cock washers. These tiny parts—found in diaphragm-type assemblies—may be causing loud tank noises or leaks. Remove the two retaining pins atop the ball cock assembly that hold the float arm in place. Lift the float assembly out of the tank.

With pliers, pull the plunger up and out of the ball cock. Inside the plunger (see **Drawing 61**) you'll find washers: a seat washer and one or more split washers. If they're worn, replace them with exact duplicates. If the ball cock still leaks, replace the entire assembly.

Removing the ball cock assembly. With an adjustable wrench, unfasten the coupling and gasket that connect the water-inlet pipe to the underside of the tank. Examine this coupling and replace it if it's worn. Remove the float ball and arm inside the tank. Using locking grip pliers and a wrench, unfasten the locknut and washer that hold the ball cock shaft to the tank (see **Drawing 62**). Lift out the old ball cock and replace it.

Installing a new ball cock assembly. Put a washer and locknut—in that order—on the bottom of the new assembly. Secure the assembly by tightening the nut. Install and tighten the coupling and gasket on the inlet pipe under the tank. Place the bowl-refill tube in the overflow tube (see **Drawing 63**). Attach the float arm and ball to the ball cock. Tighten the coupling and locknut firmly.

Turn on the water and allow the tank to fill. Adjust the float arm.

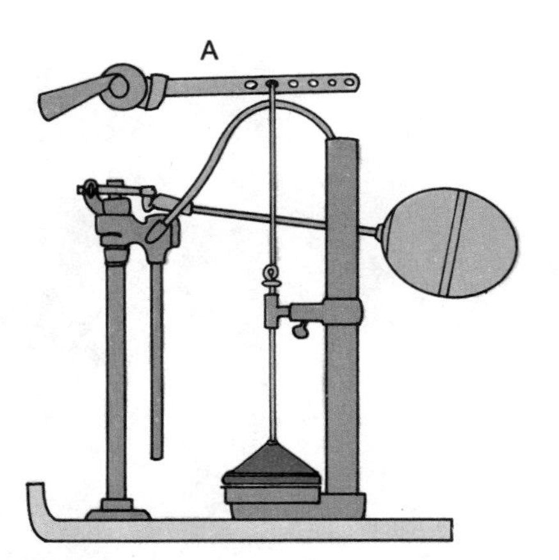

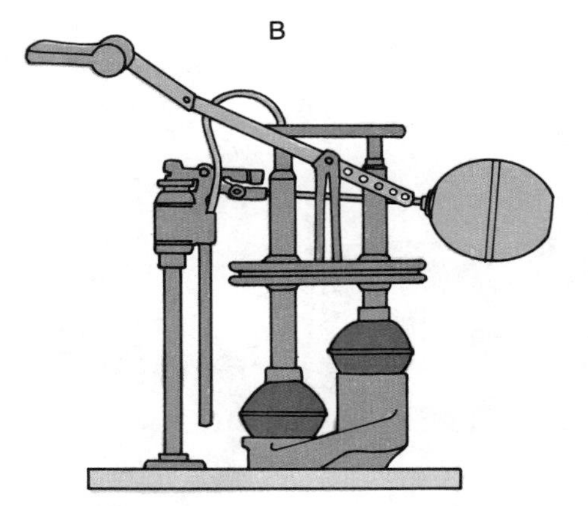

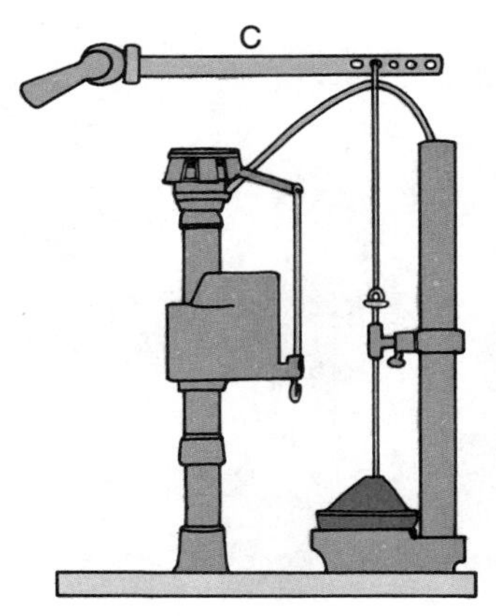

Drawing 60. Ball cock assemblies: (A) diaphragm-type, (B) two-way flush valve, (C) float-cup ball cock.

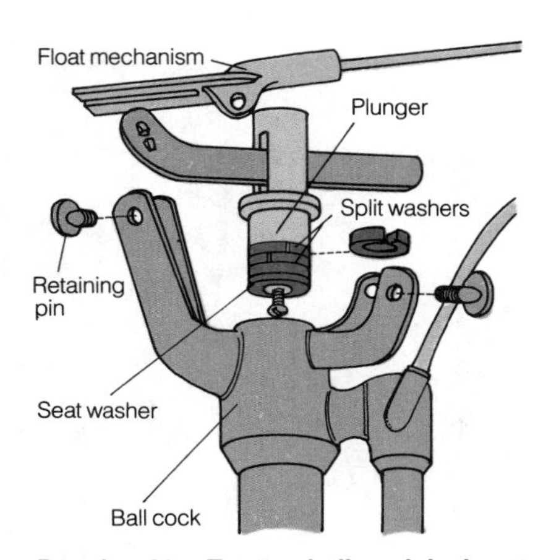

Drawing 61. To stop ball cock leaks, remove the plunger from the ball cock and replace defective washers.

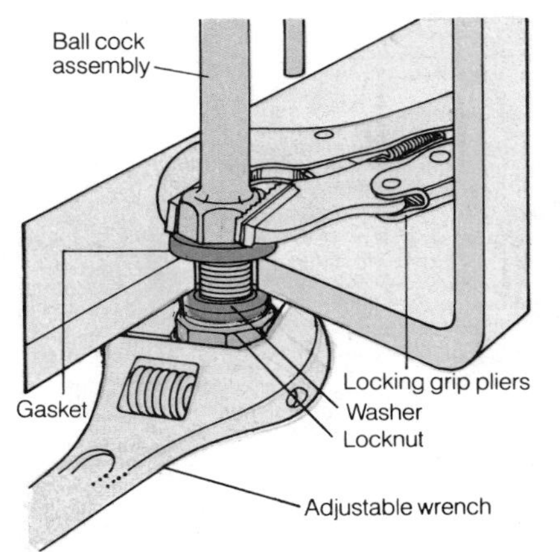

Drawing 62. To remove the ball cock assembly, use the counterforce of pliers and a wrench to detach the locknut.

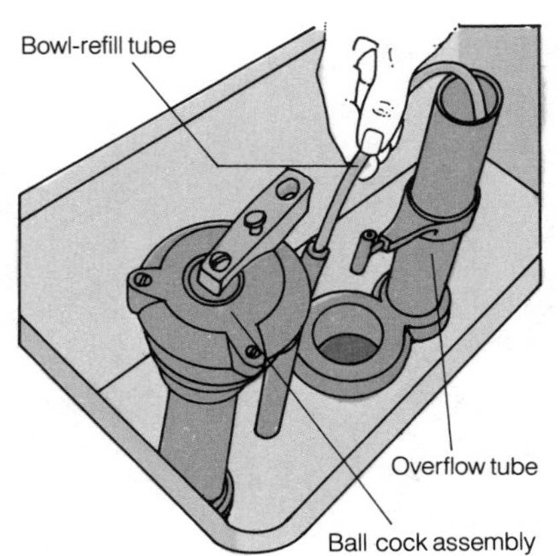

Drawing 63. To install a ball cock assembly, position the bowl-refill tube in the overflow pipe.

Running Toilets

Ready to put an end to that incessant trickling and gurgling in the tank? Refer to the chart on page 32 for help in diagnosing the exact cause of your toilet's running; Drawing 59 on page 32 will help you pinpoint the problem area. Then try one of these repairs.

CAUTION: Before working on the stopper or the valve seat, turn off the water at the fixture valve or the main shutoff valve (see page 23).

Adjusting the float arm. One solution to a running tank is bending the float arm downward (see **Drawing 64**) to seal the flush valve. Also, check to see if the ball is sticking against the back of the tank. If it is, adjust it. Be sure to use both hands, and work carefully to avoid straining the assembly.

Replacing the float ball. The float ball, which can be either plastic or metal, sometimes becomes perforated and fills with water. If this happens, unscrew the ball and replace it.

Working on the stopper and valve seat. A common cause of a continuously running toilet is a defective tank stopper (see **Drawing 65**) or valve seat. The two parts may not make a tight seal if the seat is uneven or the stopper is off center. To check, remove the lid and flush the toilet to empty the water. Watch the stopper; it should fall straight down to stop the flow of water from the tank. If it doesn't, you'll have to make an adjustment as described in "Installing a flapper stopper," at right.

Next, inspect the valve seat (usually made of brass, copper, or plastic) for corrosion or mineral buildup. If you find a problem, gently scour with fine steel wool (see **Drawing 66**).

Examine the stopper. If it's soft or out of shape, replace it with a new one (directions follow). If the water still runs, you'll need to replace the flush valve assembly (directions follow) or repair the ball cock (see page 33).

Installing a flapper stopper. If your tank stopper needs replacing, install the flapper type with a chain to eliminate any future misalignment problems with the lift rod wires or guide arm (see **Drawing 67**). It's an easy replacement. Flush to drain the tank. Unhook old lift wires from the trip lever, then lift out guide rod and wires. Slip

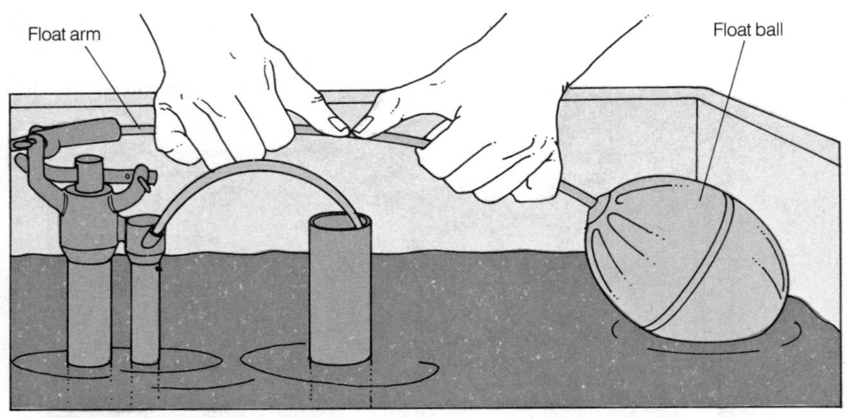

Drawing 64. Bending the float arm downward lowers the water level in the tank so the toilet doesn't continue to run.

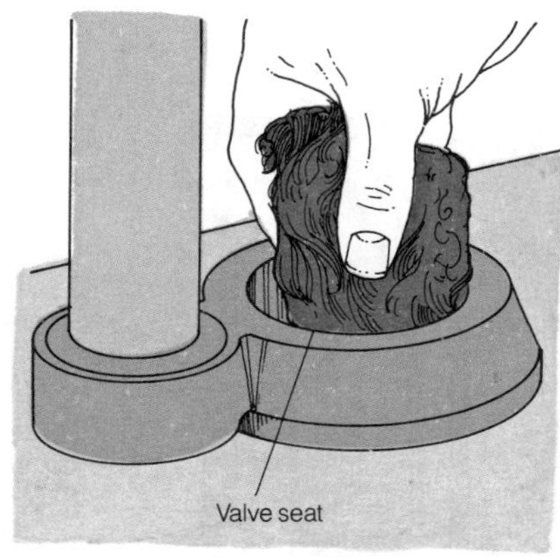

Drawing 66. To clean the flush valve seat, gently scour the seat, rim, and outlet valve with fine steel wool.

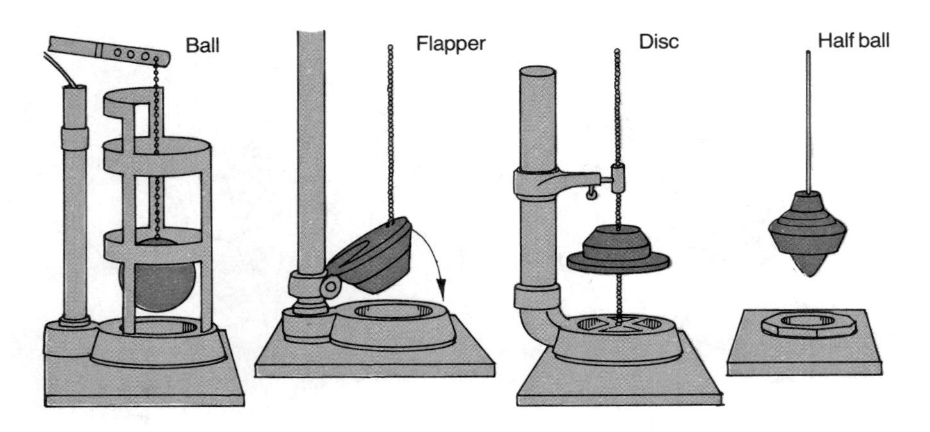

Drawing 65. Tank stoppers: ball stopper, flapper, disc, half ball.

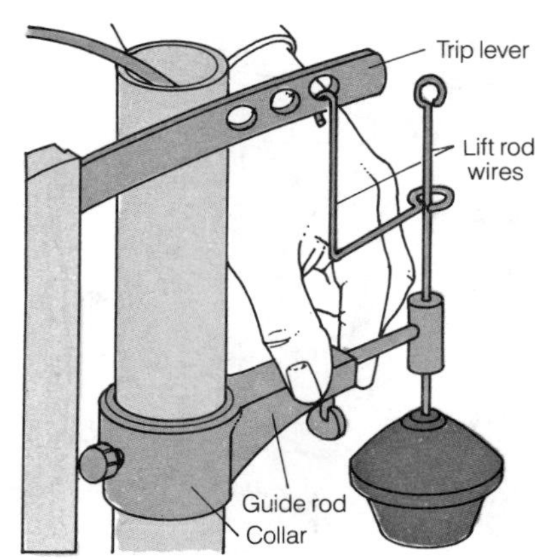

Drawing 67. To install a flapper stopper, first remove the upper and lower lift rod wires and the guide rod.

the new flapper down over the collar of the overflow tube and fasten the chain to the trip lever.

Replacing the flush valve assembly. First, empty the tank by flushing. Remove the old stopper, guide rod, lift wires, or chain.

If you have an older wall-hung toilet (see **Drawing 68**), loosen the coupling on the short 90-degree pipe bend under the tank and remove the pipe. Unscrew the locknut on the discharge tube; then remove the valve seat and the gasket.

For a bowl-mounted tank (see **Drawing 69**), remove the bolts, then the gaskets.

Insert the discharge tube of the new valve assembly through the tank bottom. Position the overflow tube (see **Drawing 70**) and tighten the locknut to hold it in place. Center the guide rod on the overflow tube (see **Drawing 71**) over the valve seat and tighten it in place. Install the lift wires through the guide rod and trip lever (page 32). Screw the stopper onto the lower lift wire (see **Drawing 72**), aligning it with the center of the seat.

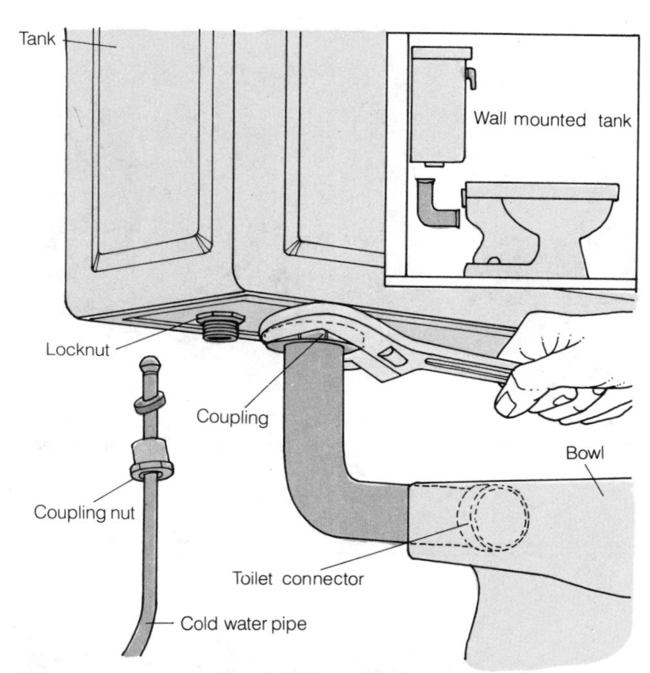

Drawing 68. Wall-hung toilet has a bend of pipe that connects the tank to the bowl.

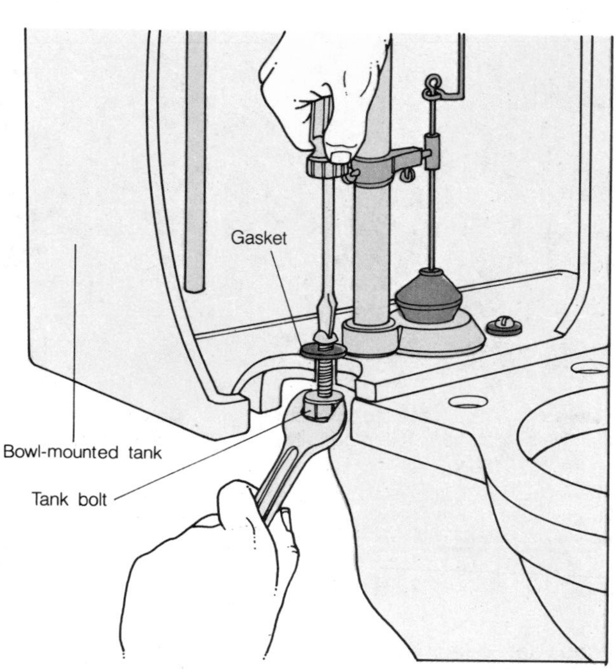

Drawing 69. Bowl-mounted tank toilet has two bolts that fasten the tank atop the bowl.

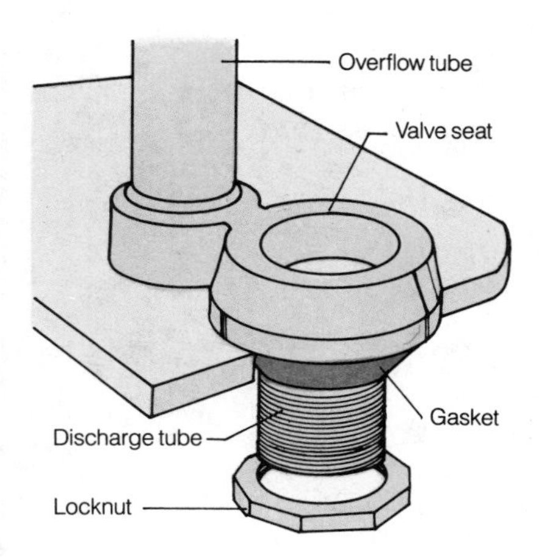

Drawing 70. Position the overflow tube towards the ball cock assembly and tighten the locknut that holds the discharge tube.

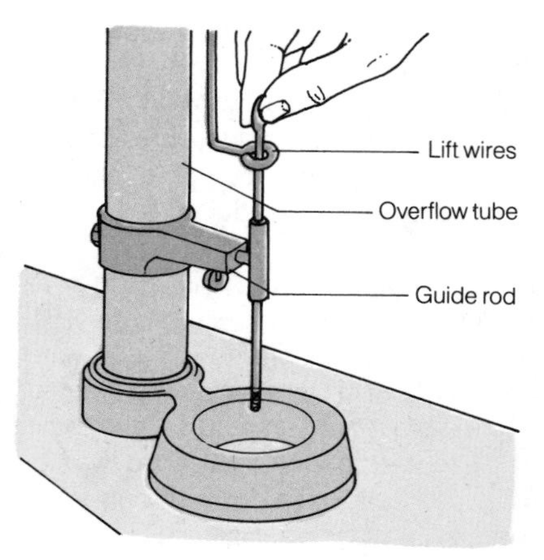

Drawing 71. Center the guide rod on the overflow tube and install the lift wires above the valve seat.

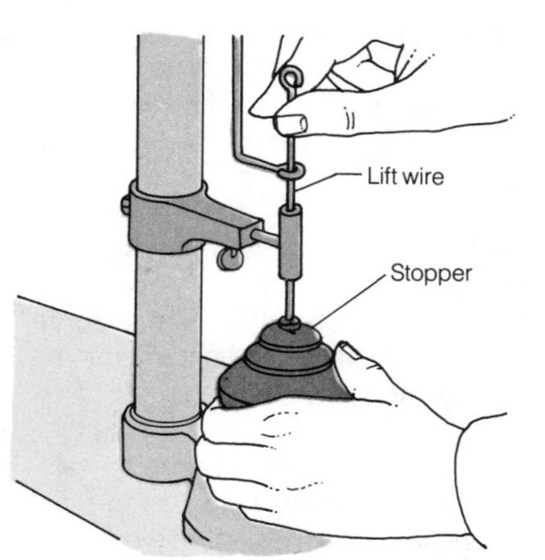

Drawing 72. Screw the stopper onto the lower lift wire, positioned directly above the valve seat.

Clogged Toilets

If you suspect a toilet is clogged, don't flush it or you'll have a flood on the floor. But if you see an overflow about to happen, you can usually prevent it by quickly removing the tank lid and closing the flush valve by hand (see **Drawing 73**). Just push the stopper into the valve seat.

The usual cause of a clogged toilet is an obstruction in the trap. To remove it, use a plunger; if that doesn't work, use a closet auger. If these don't clear the clog, you might try using a hose or balloon bag (see page 30) in the nearest cleanout.

CAUTION: Don't use any drain cleaners in a toilet. Not only are they ineffective, but they can harm the porcelain and present the additional problem of caustic water.

Using a plunger. First try plunging the clog. Use a funnel-cup plunger—it has a special tip to fit the bowl (see **Drawing 74**). Pump it up and down a dozen times to push the obstruction through the trap or draw it back into the bowl.

Using a closet auger. If the plunger doesn't clear the clog, the next step is to use a closet auger (see **Drawing 75**). This tool will reach down into the toilet trap. It has a curved tip that negotiates curves with a minimum of mess, and a protective housing to prevent scratching the bowl. Follow the directions on page 27 for using a snake.

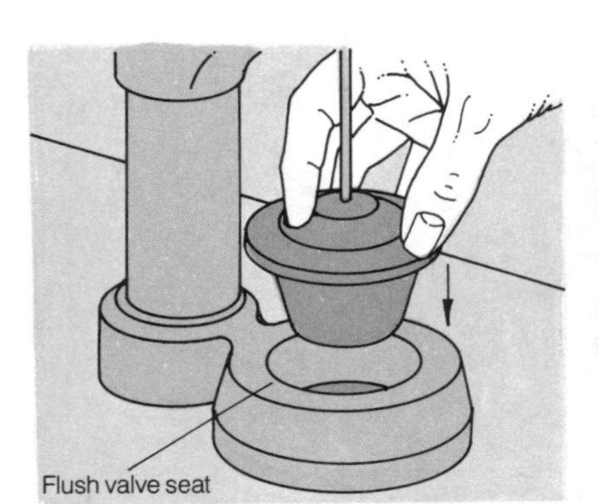

Drawing 73. Push the stopper into the valve seat by hand to temporarily prevent a toilet overflow.

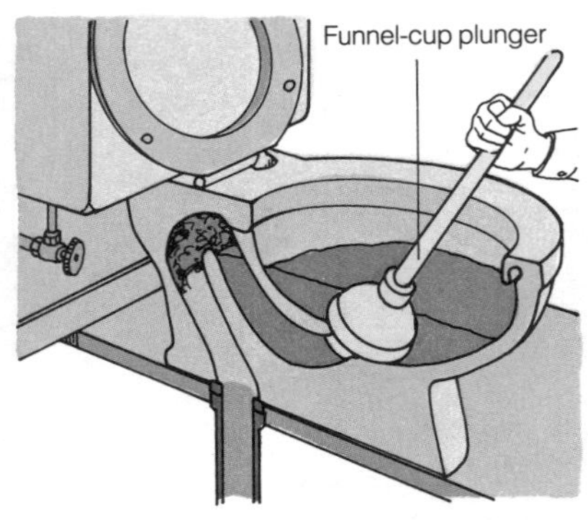

Drawing 74. Use a funnel-cup plunger, specially designed for toilets, to dislodge a clog in the toilet trap.

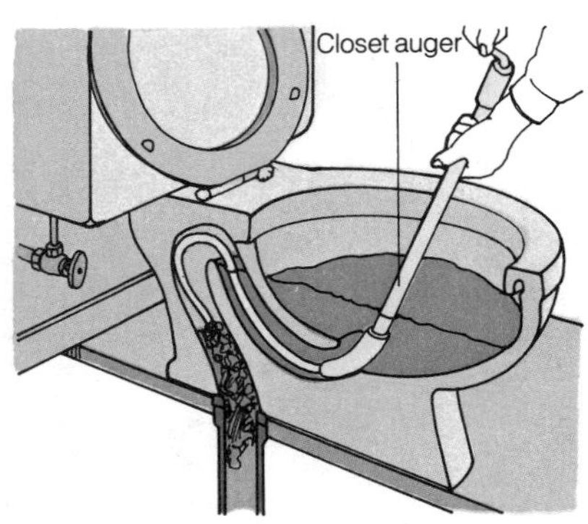

Drawing 75. Use a curve-tipped closet auger to reach a deep-set blockage in the toilet trap.

Septic tank maintenance

A good septic tank system doesn't require a great deal of maintenance or call for many special precautions. But the maintenance it does require is crucial, since failure of the system can constitute a serious health hazard. You should have a diagram of your septic tank's layout, showing the location of the tank, pipes, manholes, and drainage field.

Chemicals, chemical cleaners, and thick paper products should never be disposed of through the system. Some chemicals destroy the bacteria necessary to attack and disintegrate solid wastes in the septic tank. Paper products can clog the main drain to the tank and smaller pipes to the dispersal field, making the entire system useless.

Have your septic tank checked at least once a year. To function properly, the tank must maintain a balance of sludge (solids remaining on the bottom), liquid, and scum (gas containing small solid particles). The proportion of the sludge and scum layers to the liquid layer (see **Drawing A**) determines whether pumping is needed.

Inspecting and pumping should be done by professionals.

Have your septic tank pumped whenever necessary, but plan ahead if you can. It's best to remove the sludge and a portion of the scum in the spring. If you have it done in autumn, the tank will become loaded with solid waste that can't be broken down through the winter, when bacterial action is slowed.

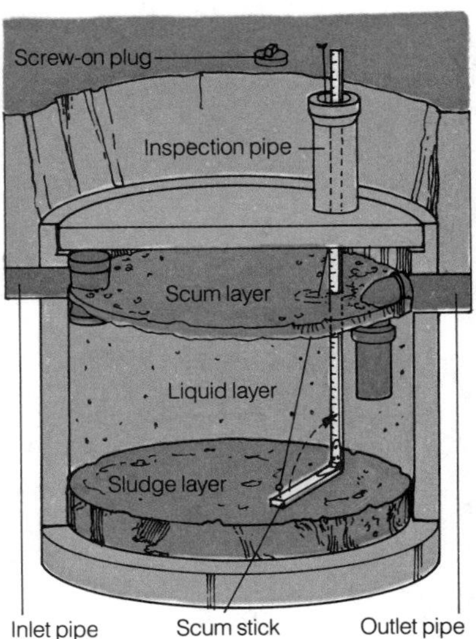

Drawing A. Sludge, liquid, and scum layers in septic tank.

Toilet Sweating, Leaks & Flush Problems

Miscellaneous toilet problems you may need to repair include sweating (moist air condensing on the tank), outright leaks, and difficulties with the handle or the flush passages.

CAUTION: Before working on a sweating tank problem or a leak, turn off the water at the fixture valve or main shutoff valve (see page 23), and empty and sponge out the tank.

Tank sweating. A common problem, tank sweating happens more often in summer, when cold water in the tank cools the porcelain and warm, moist air condenses on the outside. More than an inconvenience, sweating encourages mildew, loosens floor tiles, and rots subflooring.

An easy solution: Insulate the inside of the tank with a special liner sold at plumbing supply stores, or create a liner with foam rubber.

To make your own tank liner, cut pieces of ½-inch-thick foam rubber to fit inside the tank. Apply a liberal coating of silicone glue or rubber cement to the tank sides and press the foam in place (see **Drawing 76**). Let the glue dry for 24 hours before refilling the tank. Make sure the pads don't interfere with any moving parts.

If the water entering the tank is often below 50°, you may need a tempering valve (see **Drawing 77**) that mixes hot water with cold. A tempering valve requires a hot water hookup; installation is neither cheap nor easy.

Leaks. To stop a leak between the tank and bowl of a bowl-mounted tank toilet (see **Drawing 78**), you'll need to remove the tank (see page 35) and replace the gasket. If the leak persists, check to be sure the flush valve assembly isn't defective (see page 35).

If the leak is at the water-inlet pipe, disconnect the coupling under the tank and check its washer for wear.

If the base of the toilet bowl is leaking, you'll need to replace the wax gasket that seals the bowl to the floor flange. Begin by removing the tank (see page 35) and bowl (see pages 78–79). Then, with a putty knife, remove the old gasket (see **Drawing 79**) from the bottom of the bowl and the floor flange. Check the floor flange for deterioration; replace it if it's defective.

Place a new wax gasket (see **Drawing 80**) on the toilet opening (called the horn). Apply plumber's putty around the bowl's bottom edge. Replace the toilet (pages 78–79).

Flush problems. A loose handle or trip lever can cause an inadequate or erratic flush cycle. Tighten the setscrew on the handle linkage (see **Drawing 81**) or replace the handle.

Use a piece of wire to unclog flush passages under the bowl's rim.

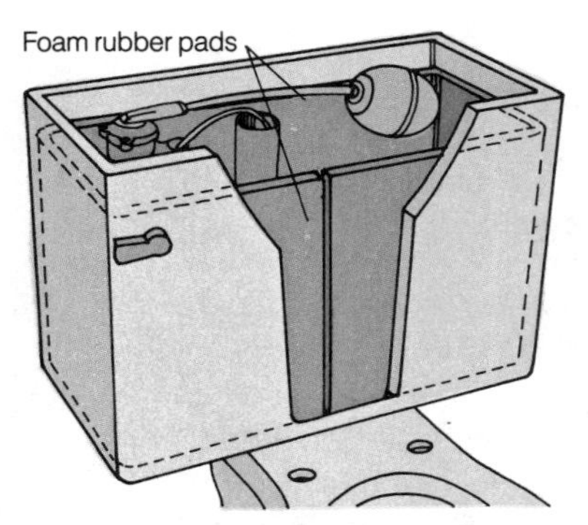

Drawing 76. Install a foam tank liner to insulate a toilet tank and prevent tank sweating.

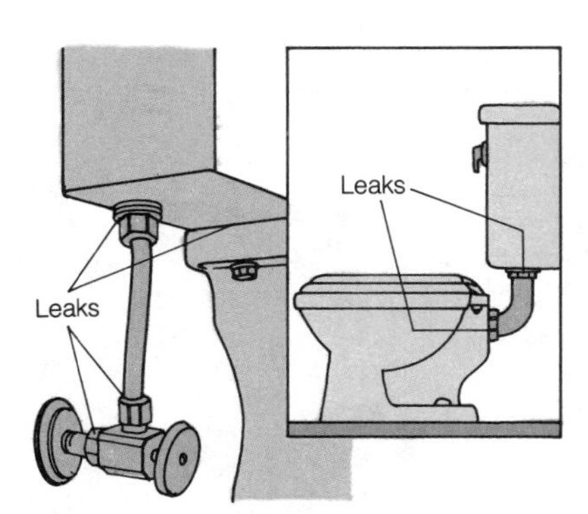

Drawing 78. Repair leaks between the bowl, tank, and wall by replacing connecting gaskets and washers.

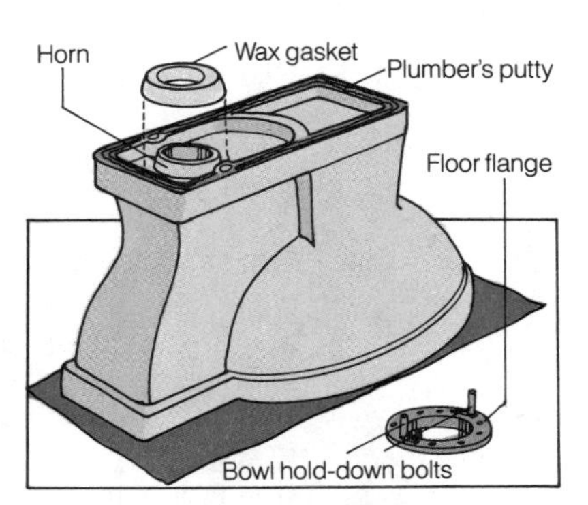

Drawing 80. Install a new wax gasket on the toilet horn to make a watertight seal with the floor flange.

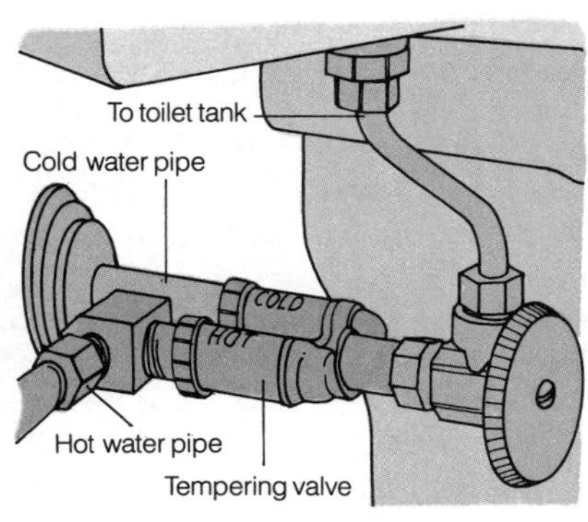

Drawing 77. A tempering valve mixes hot water with cold to stop troublesome tank condensation.

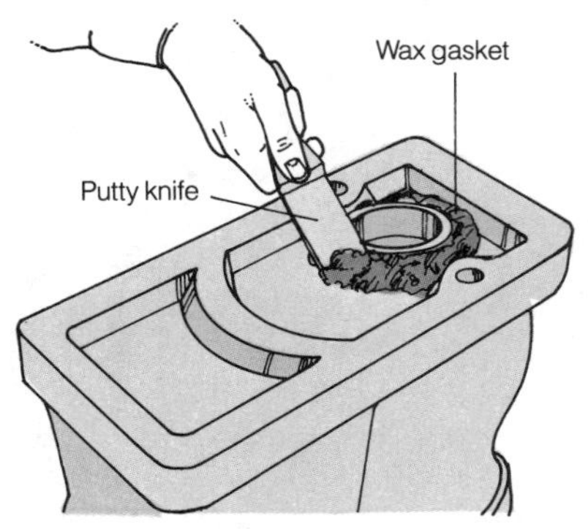

Drawing 79. Replace the wax gasket beneath the bowl to stop leaks at the base of the toilet.

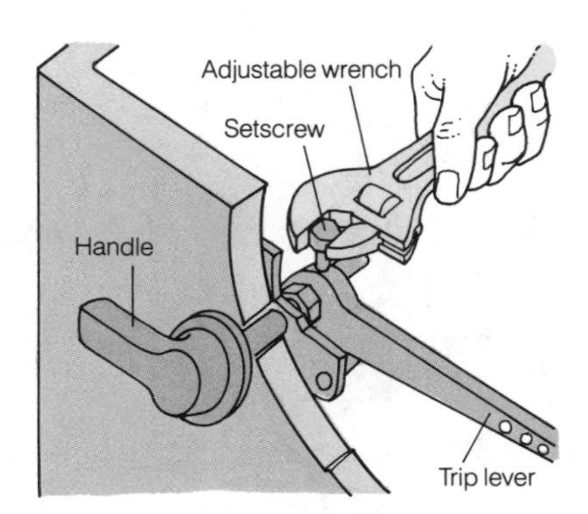

Drawing 81. Tighten the setscrew on the handle or trip lever if a loose handle is causing an inadequate flush.

Water Heater Problems

Considering how much we depend on them, it's fortunate that water heaters are among the longer-lasting appliances. Relatively simple in design, conventional water heaters are commonly gas-fired (see **Drawing 82**) or electric (see **Drawing 83**) units. (Recent developments include tankless instantaneous water heaters that use up to 20 percent less fuel than the tank models.

In a conventional heater, cold water is carried to the tank bottom through the dip tube; hot water leaves the top of the tank as it is used. When the water temperature lowers, the thermostat activates the burner in the gas unit or two heating elements in the electric model. In gas (and oil) heaters, the flue, which runs up the center, vents harmful gases to the outside; electric heaters don't need venting.

Each heater has a drain valve at the tank bottom to draw off water, sediment, or rust from inside the tank. Open the drain valve once every few months to drain out any cloudy water. Also, test the temperature and pressure relief valve on the top of the tank as a safety measure (see "A malfunctioning relief valve," following).

Gas water heaters. Be sure you know how to light the pilot light (see **Drawing 82**). The heater should come with a handbook and should also have lighting instructions printed on it near the thermostat. It's very important to have the flue connected properly so combustion residue will escape harmlessly. For safety reasons, never make repairs on the gas burner yourself.

Electric water heaters. The thermostat is usually concealed behind a metal door on the side of the tank (see **Drawing 83**). If you need to disconnect the wiring to the heater, be sure to shut off the power at the fuse box or circuit breaker before working.

A malfunctioning relief valve

A residential water heater has a temperature and pressure relief valve installed in the top of the tank. The valve allows steam or water to escape safely in case of a malfunction in the thermostat or a pressure buildup.

If the temperature and pressure relief valve frequently opens to let out steam or water, shut off the heater and check with the utility company. The temperature and pressure relief valve may be defective, or there may be a buildup of temperature and pressure. It's an easy job to replace an ailing valve. But replacement of any of the components responsible for excessive pressure and temperature is a job for professionals.

The temperature and pressure relief valve should be tested periodically (see **Drawing 84**). If there is no overflow pipe connected to the relief valve, attach a hose to the valve and run the hose to a pail. Lift the valve handle—hot water should drain into the hose or pipe. If not, install a new valve.

A leaking tank

Should you discover that your water heater tank has a hole in it, you can make a temporary repair (see **Drawing 85**), but one weak spot indicates that the rest of the tank is corroding, too. Start shopping around for a new one (see pages 84–87).

To stop the leak temporarily, first drain the water out; then remove the outer jacket and insulation. Drill the leaking hole just large enough to accommodate a toggle bolt, then insert the bolt and tighten it to stop the leak.

No hot water

In a gas water heater, a lack of hot water can usually be traced to one of the following circumstances, and easily remedied:

- *The pilot light is extinguished.* Clean dust and lint from the area and relight the pilot according to instructions on the tank.
- *The gas inlet valve on the supply pipe has been inadvertently closed* (turned so the handle is at right angles to the pipe). Open it so that the

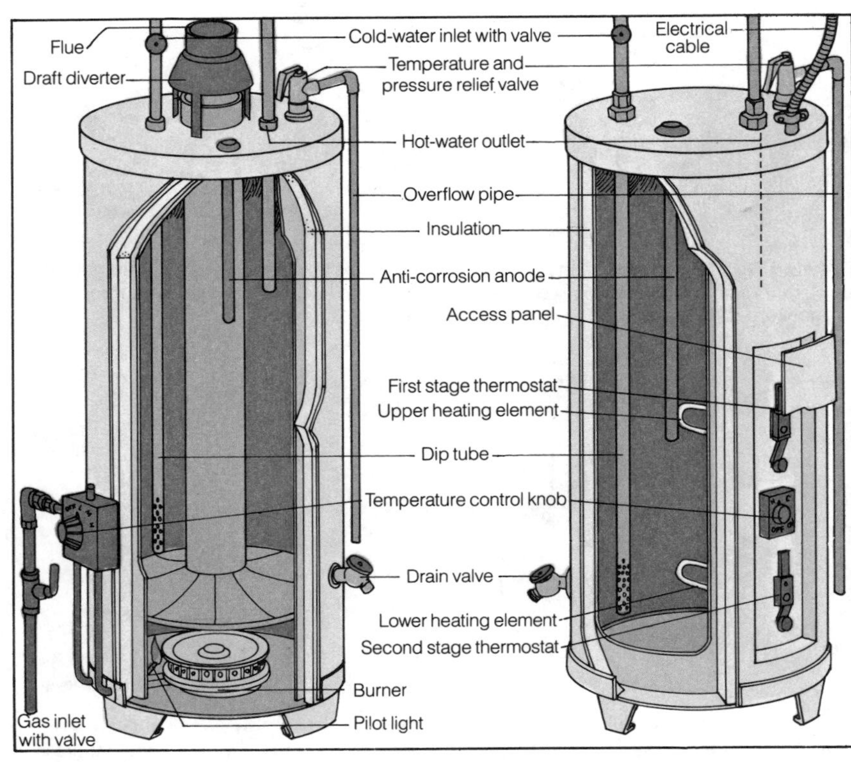

Drawing 82. Gas water heater

Drawing 83. Electric water heater

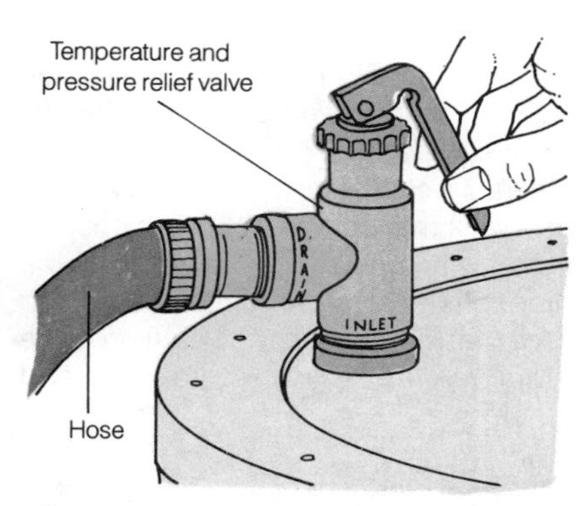

Drawing 84. To test a temperature and pressure relief valve, lift the valve handle until hot water drains out.

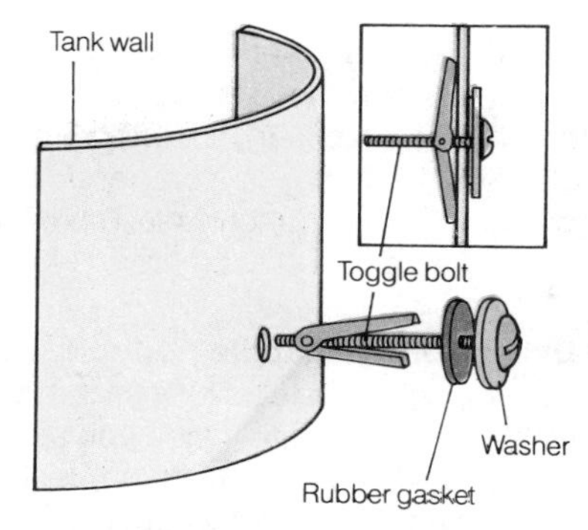

Drawing 85. To make a temporary tank repair, drain the tank and drill the leaking hole large enough to insert a toggle bolt.

handle is parallel to the pipe. Then relight the pilot according to instructions on the tank.

- *The thermostat has been shut off or is defective.* Turn the thermostat switch to the ON position. If that doesn't solve the problem, replace the thermostat.

In an electric water heater, check the following causes of no hot water:

- *The heater switch has been shut off.* Turn it to ON.
- *The fuse has blown or the circuit breaker supplying the heater has tripped.* Restore power.
- *The upper heating element is burned out or has a calcium buildup, or the thermostat is defective.* Have a technician check these out and make necessary replacements.

Insufficient hot water or water not hot enough

Whether gas or electric, a water heater may not hold enough water to meet your needs. (See the chart on page 84 that explains how to choose a water heater sized for your family's needs.) Small heaters need as much as a half-hour to recover after a full bath or a washing machine cycle. Before you decide on a larger heater, though, check the following possibilities.

If you have a gas water heater:

- *The thermostat dial is set too low.* If the thermostat is in view, adjust it upward. The setting should give temperatures up to 160°—the heat necessary to run a dishwasher efficiently. If you can't find the thermostat, it's probably buried in the insulating layer around the tank, and adjusting it will require a visit by the utility company. Do not disturb the insulation yourself.
- *The gas inlet valve on the supply pipe is partially closed.* Open it fully—so the handle parallels the pipe exactly.

If you have an electric water heater:

- *The thermostat is set too low or is defective.* Call the utility company. To work properly, the thermostat must be buried in the insulation, so it is out of reach. If you disturb the insulation, you're almost certain to make the thermostat function less accurately than before.
- *The lower heating element has burned out.* Call a technician to replace it. The utility company's service department does not make repairs, only adjustments to the heating equipment.

Water too hot

If you have a gas water heater:

- *The thermostat is set too high.* Adjust it to a lower setting. But remember, 150° to 160° is recommended if you have a dishwasher. If steam or boiling water comes from a tap, the thermostat has failed and must be replaced. If this happens, shut off the gas inlet valve to the heater and call for professional help.
- *The burner doesn't turn off.* When the thermostat fails to shut off the burner, you may see steam coming from the faucets. Shut off the gas inlet valve to the heater until you can get professional help.

If you have an electric water heater:

- *The thermostat is set too high.* Adjust to a lower setting. But remember, 150° to 160° is recommended if you have a dishwasher. If steam or boiling water comes from a tap, the thermostat is malfunctioning. Suspect a short or burned terminals—problems that should be dealt with by a professional.
- *Thermostat heating element or high-temperature cutoff is defective.* Call for repairs; professional help is needed.

Noisy water heater

Causes—and cures—of water heater noise are the same whether the heater is gas or electric.

- *A buildup of sediment in the tank.* You'll hear loud rumbling and cracking sounds if there's sediment in your tank. As sediment accumulates, small quantities of water get trapped in its layers. When this trapped water gets very hot, it explodes out of its pocket. The rumbling and cracking noises are actually a series of tiny steam explosions that can be annoying and harmful.

  Attach a hose to the drain valve at the tank's base. Leaving the cold water supply turned on, open the drain valve and flush the tank for about 5 minutes. The forceful flow of incoming water should help wash away sediment. When the discharging water becomes clear, close the valve and remove the hose.
- *A buildup of steam in the tank.* Check the temperature and pressure relief valve. (See "A malfunctioning relief valve" on the facing page.) The valve should be periodically releasing steam. If it does not, you may need to replace it.

Garbage Disposer Breakdowns

If your garbage disposer makes a loud whirring noise or stops functioning entirely, it has jammed.

CAUTION: Do not attempt to clear a clogged disposer drainpipe with chemicals of any type. They probably won't work, and you'll have the added problem of a pipe filled with caustic solution. For safety, never put your hand in a disposer—try to improvise another way to remove anything that may fall into the disposer. Pliers or kitchen tongs are good possibilities.

Resetting the mechanism. Disposers have built-in safeguards. Some are equipped with an automatic shutoff that protects both the house wiring and the appliance when jamming occurs. Another type of cut-out switch stops only the grinding mechanism.

If the disposer jams, turn the switch off (or remove the cover, if it's a batch-feed disposer), wait 3 to 5 minutes for the motor to cool, and then press the reset button (see **Drawing 86**) at the bottom of the motor.

Using a broom handle. You can often cure disposer jams quickly and easily with a broom handle. Once you make sure the disposer is off, angle the broom handle (or a 1 by 2-inch board) against one of the impeller blades (see **Drawing 87**) and pry. Work the pivoting flywheel back and forth until the jam dislodges. Clear out the blockage and push the reset button to restart the motor.

Reaching from below. Instead of using a broom handle (or if the broom handle technique doesn't work), you can try to fix the disposer from below. Beneath some garbage disposers there's a socket for an Allen wrench (or a crank that comes with the disposer) that will turn the flywheel. Deluxe models have an automatic reversing mechanism that allows you to free the jam by flipping a switch off and then on.

Using a snake. If your disposer drainpipe clogs, disassemble the trap (see pages 24–25) and thread a snake into the drainpipe (see **Drawing 88**).

If both basins of a double sink are clogged, snake down from the one without a disposer. If only the basin with the disposer is clogged, you'll have to remove the trap to dislodge the blockage.

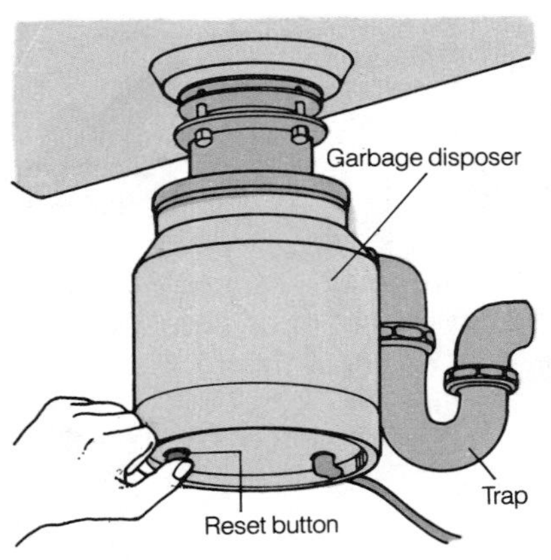

Drawing 86. Push the reset button before reactivating a garbage disposer.

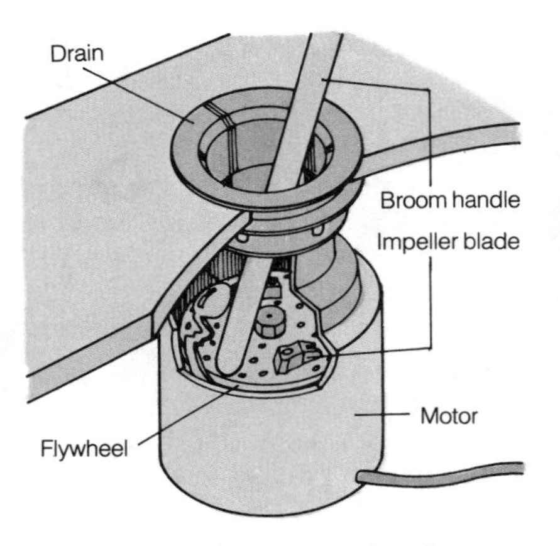

Drawing 87. Use a broom handle to dislodge material that's jamming the flywheel.

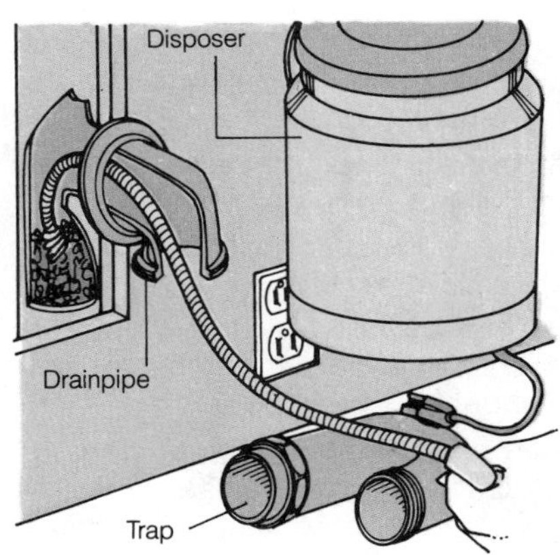

Drawing 88. Clear a clogged drainpipe by maneuvering with a snake.

Preventive care for garbage disposers

Generally, garbage disposers are tough—they'll gobble up just about anything you serve them. The most expensive models can even handle bones, corncobs, and seafood shells without jamming (a kind of mechanical indigestion). Protect your disposer's good health by taking the following precautions.

- *Keep these out of the disposer:* bottle caps, cans, glass, rags, string, paper, rubber, plastic, foam, or cigarette filters. Check your owner's manual for kinds of food to avoid, too. If you're unsure about putting a substance in the disposer, throw it in the garbage instead.
- *Dispose of moderate amounts at a time*—overloading causes jams.
- *Always use cold water* (and lots of it) to wash away ground-up waste. Keep water running whenever the disposer is turned on. Cold water congeals grease so it flushes away instead of lodging in the pipes.
- *Deodorize the disposer* by filling it three-quarters full of ice cubes. Turn the unit on (expect a loud grinding sound) and flush with cold water. Finally, drop half a lemon into the disposer and turn it on.
- *Flush the disposer* with a solution of 1 quart hot water and ½ cup washing soda (a laundry product sold in supermarkets). Repeat the treatment as needed.

Dishwasher Dilemmas

Automatic dishwashers are invaluable kitchen helpers. But like all machines, they're fallible. Most of the dishwasher-related plumbing problems you'll encounter involve water-inlet valves or drains. Many of these plumbing problems are easy enough to tackle yourself. But if your dishwasher has a mechanical problem, such as a defective solenoid, it's best to call in a technician to do the work.

Remedies for common dishwasher problems. The chart that follows gives causes for most of the problems you'll encounter with your dishwasher. Take a close look at our typical dishwasher (see **Drawing 89**) and then at your own dishwasher to pinpoint the exact cause of the problem.

Air gap maintenance. Many plumbing codes require that an air gap (see **Drawing 90**) be installed on dishwashers to prevent siphoning of wastewater from the sink drain or garbage disposer back into the washer. (If your dishwasher has no air gap, it has a high loop in its drain hose instead.) The air gap is a chrome cap or pipe on the sink or countertop.

Dishwasher water discharges through a drain hose connected to one of the two pipes on the bottom of the air gap. Another drain hose runs from the air gap to the sink trap or disposer.

Because bones, seeds, or bits of food from the dishwasher or disposer can lodge in the air gap, you should clean it out periodically. Lift off the cap, unscrew its cover, and insert a long piece of wire, pushing it straight down. This may also help if the dishwasher is slow to drain, or if suds ooze from the cap.

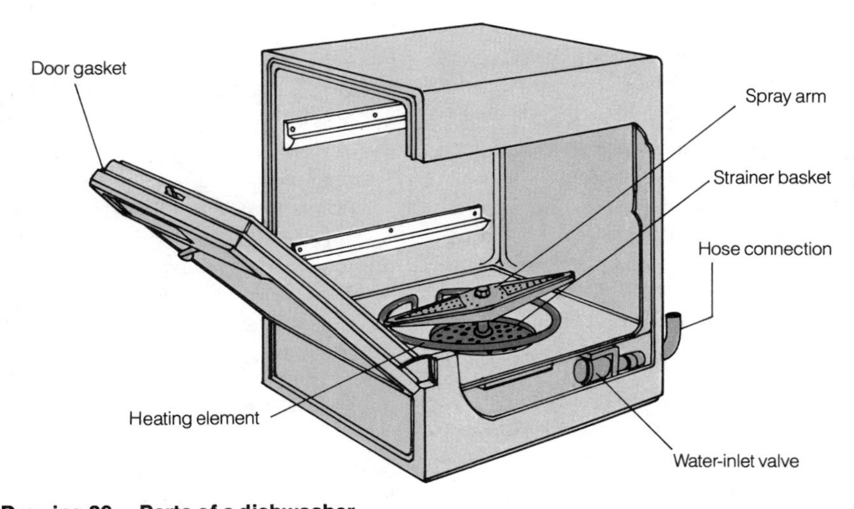

Drawing 89. Parts of a dishwasher

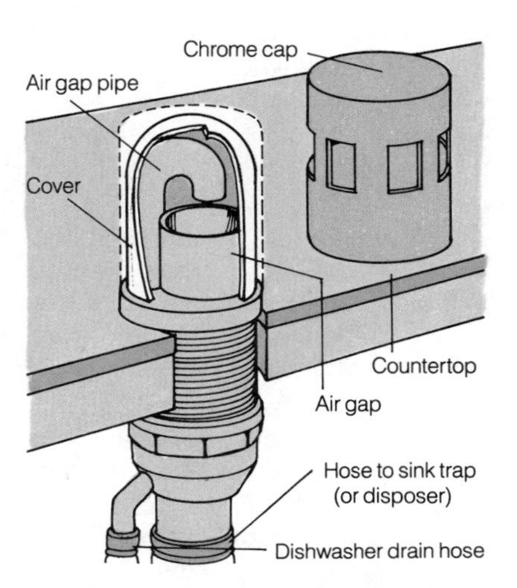

Drawing 90. Air gap works like a high loop in the drain line to prevent siphoning.

Problem	Causes	Remedies
Dishwasher won't drain	Plugged strainer basket	Clean dirt, grease, food buildup from strainer in bottom of tub
	Clogged drain	Open trap cleanout under sink
	Dirty air gap	Clean out air gap on sink or countertop (see "Air gap maintenance," above)
Dishwasher leaks	Faulty hose connection	Tighten or replace hose (turn off electricity first)
	Door gasket isn't sealing	Tighten loose gasket; replace torn or split gasket
Dishwasher continues to fill	Faulty solenoid	Have solenoid repaired or replaced
	Water-inlet valve won't close	Disassemble and clean valve parts
	Blocked fill spout in water-inlet valve	Disassemble water-inlet valve and clean fill spout
Dishwasher doesn't clean dishes	Low water temperature	Turn up water heater to 150° to 160° range (160° is ideal)
	Low water pressure	Call water company
	Jammed spray arm	Check to see if something is obstructing spray arm
	Clogged spray arm holes	Clean residue buildup from spray arm holes
Dishwasher won't fill	Closed water-inlet valve	Open water-inlet valve on supply line
	Water-inlet screen blocked	Clean calcium buildup off screen or replace water-inlet valve
	Faulty solenoid	Have solenoid repaired or replaced
	Low water pressure	Call water company

The Highs & Lows of Water Pressure

Most appliances, valves, and fixtures that use water are engineered to take 50 to 60 psi. Mains deliver water at pressures as high as 150 psi and as low as 10 psi. Too much pressure is much simpler to cure than too little pressure.

Low pressure

The symptom of low pressure is a very thin trickle of water from faucets throughout the house. Chronic low pressure is typically found in homes on hills near reservoir level or in old homes whose pipes are badly clogged by scale and rust due to hard water. In many communities, periodic low pressure may occur during peak service hours through no fault of the home's location or plumbing.

Remember that fixtures farther from the main water supply pipe get less water pressure than those nearer the main. Too-small pipes may aggravate the problem of low pressure, too.

If you add new fixtures, you may need to install a larger main supply pipe from the point where the water enters the house to the various branches of the supply network, in order to maintain satisfactory pressure.

Whatever the cause of chronic low pressure, you'll probably want to think of improvements rather than complete cures. The latter are either very expensive or mechanically unfeasible—they range from complete replacement of the plumbing to building your own reservoir in a tower tank above the house. Still, you can make modest improvements by flushing the pipes and by replacing the main supply pipe.

Flushing the pipes. A system showing early signs of clogged pipes can regain some lost pressure if the system is flushed. To do this, take these steps:

- *Remove and clean aerators* on faucets (see page 19).
- *Close the gate valve* that controls the pipes you intend to clean. It may be a shutoff valve on the water heater, or the main shutoff valve (see page 23).
- *Open fully the faucet* at the point farthest from the valve, and open a second faucet nearer the valve.
- *Plug the faucet* near the valve (but don't shut it off) with a rag.
- *Reopen the gate valve* and let water run full force through the farther faucet for as long as sediment continues to appear—probably only a few minutes. Close the faucets, remove the rag, and replace the aerators.

Replacing the main supply pipe. Before you decide to replace all the pipes in your house because of low pressure, try replacing the one section that leads from the main shutoff valve to your house (see pages 7 and 9). If it's a ¾-inch pipe, replace it with a 1-inch pipe. The larger size won't increase the pressure, but the increased volume of water will compensate. And if you have a water meter, you can also ask your utility company to install a larger one.

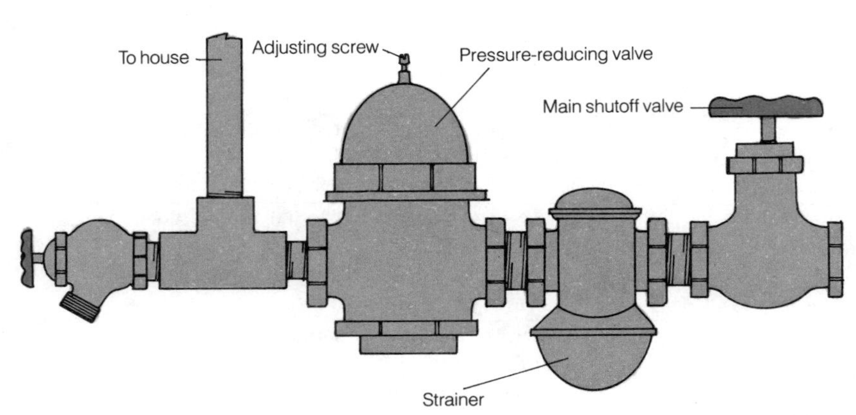

Drawing 91. Install a pressure-reducing valve between the main shutoff valve and the house to minimize a high-pressure problem.

High pressure

The symptoms of high pressure are loud clangs when the dishwasher shuts off or wild sprays when faucets are first turned on. High pressure usually occurs in houses on low-lying slopes of steep hills or in subdivisions where high pressure is maintained as a matter of fire protection.

If your house has particularly high water pressure, take precautions against appliance damage and floods.

- *Turn off the main shutoff valve* (see page 23) when you go on vacation.
- *Turn off appliance shutoff valves,* especially for a washing machine and dishwasher, when not in use.

Above-normal pressure can be cured easily and inexpensively by the installation of a pressure-reducing valve (see **Drawing 91**). This valve can reduce pipe pressure from 80 pounds per square inch (psi) or more down to a manageable 50 to 60 psi.

The method you'll use to install a pressure-reducing valve depends on the type of pipe—galvanized, copper, or plastic—in your plumbing system. First, assemble the valve with the pipe fittings necessary to connect the threads of the valve to the existing pipe. Then, after shutting off the water supply (see page 23), remove a length of pipe on the house side of the main shutoff valve long enough to accommodate the valve and the assembled fittings.

Install the valve, following the instructions for pipe fitting on pages 46–57. When the work is completed, you can turn the water back on. Be sure to check for any leaks in the new connections.

To minimize the water pressure, turn the adjusting screw at the top of the valve clockwise until the pressure is low enough to end bothersome pipe noises. Be sure the valve still supplies adequate water flow to the upper floors or to far-away fixtures in the house.

Leaking Pipes

A higher-than-normal water bill might give you the first indication of a leaking pipe. Or you might hear the sound of running water even when all fixtures are turned off.

When you suspect a leak, check first at the fixtures to make certain all the faucets are tightly closed. Then go to the water meter (see page 10), if you have one. If the dial moves, you're losing water somewhere in the system. If you don't have a water meter, you can buy a listening device that amplifies sounds when it's held up to a pipe.

Locating the leak. This isn't always easy. The sound of running water helps; if you hear it, follow it to its source.

If water stains the ceiling or drips down, the leak will probably be directly above. Occasionally, though, water may travel along a joist and then stain or drip at a point some distance from the leak. If water stains a wall, it means there's a leak in a vertical section of pipe. Any wall stain is likely to be below the actual location of the leak, and you'll probably need to remove an entire vertical section of the wall (see page 62) to find it.

Without the sound of running water and without drips or stains as tangible evidence, leaks are more difficult to find. Use a flashlight and start by checking all the pipes under the house in the crawl space or basement. It's quite likely you'll find the leak there, since leaks in other places in the house will usually make themselves apparent with water stains or dripping.

Fixing the leak. If the leak is major, turn off the water immediately, either at the fixture shutoff valve or the main shutoff valve (see page 23).

Patching a leak in a pipe is a simple task if the leak is small. The ultimate solution is to replace the pipe, but here are temporary solutions until you have time for the replacement job (see pages 47–65 for instructions on taking pipe apart).

The methods shown here for patching pipe are all effective for small leaks only.

Clamps should stop most leaks for several months if they are used with a solid rubber blanket. It's a good idea to buy a sheet of rubber, as well as some clamps, at a hardware store and keep them on hand just for this purpose.

A sleeve clamp (see **Drawing 92**) that fits the pipe diameter exactly works best. Use a rubber blanket over the leak, then screw the clamp down tight over the blanket. An adjustable hose clamp (see **Drawing 93**) in size 16 or 12 stops a pinhole leak on an average-size pipe—be sure to use a rubber blanket with the hose clamp. If nothing else is at hand, use a C-clamp (see **Drawing 94**), a small block of wood, and the rubber blanket.

If you don't have a clamp, you can still stop a pinhole leak temporarily by plugging it with a pencil point—just put the point in the hole and break it off. Then wrap three layers of plastic electrician's tape (see **Drawing 95**) extending 3 inches on either side of the leak. Overlap each turn of tape by half.

Epoxy putty (see **Drawing 96**) will stop leaks around joints where clamps won't work. The pipe must be dry for the putty to adhere, though. Turn off the water supply to the leak to let the area dry.

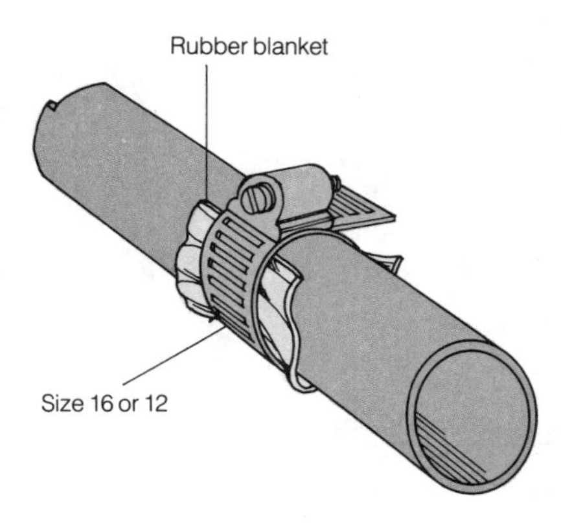

Drawing 93. Hose clamp

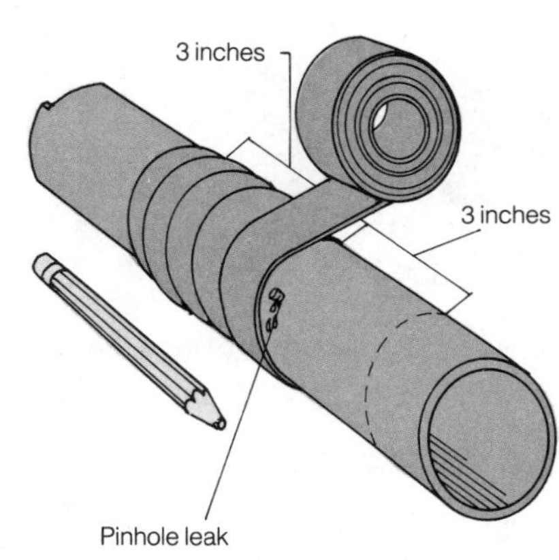

Drawing 95. Electrician's tape

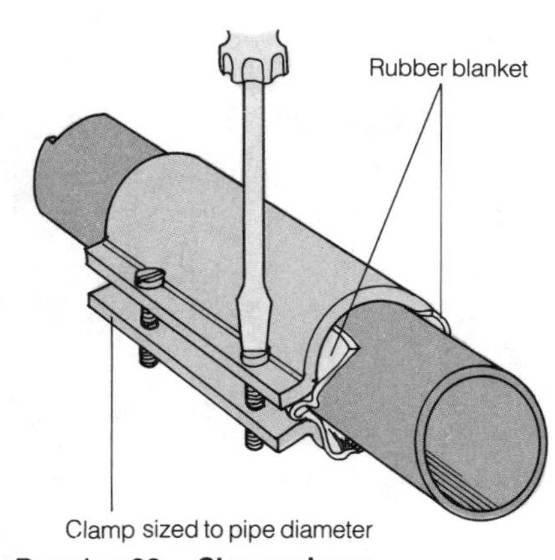

Drawing 92. Sleeve clamp

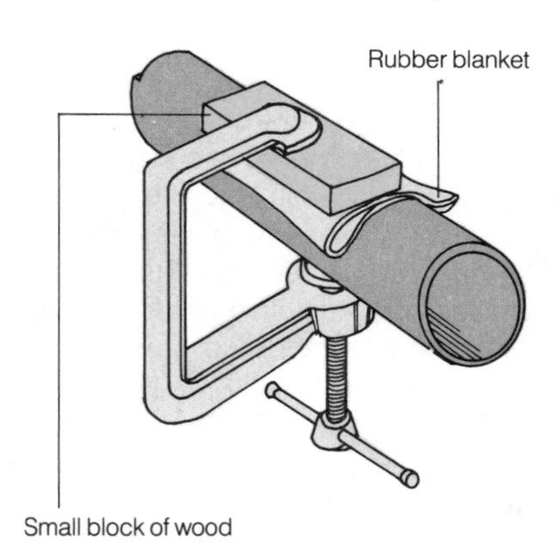

Drawing 94. C-clamp

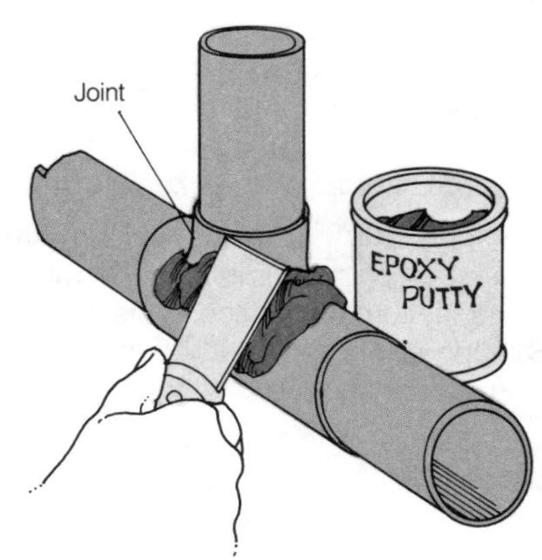

Drawing 96. Epoxy putty

Frozen Pipes

A faucet that refuses to yield water is the first sign of frozen pipes. If a severe cold snap hits, prevent freezing and subsequent bursting of pipes by following the suggestions given below. Even if the pipes do freeze, you can thaw them before they burst if you act quickly.

Preventing frozen pipes

Here's how to keep your pipes from freezing:

- *Keep a trickle of water* running from the faucets.
- *Beam a small lamp or heater* at exposed pipes.
- *Wrap uninsulated pipes* with newspapers, heating wires, foam, or tape (see page 59).
- *Keep doors ajar* between heated and unheated rooms.

Thawing frozen pipes

If a pipe freezes, first shut off the main water supply (see page 23) and open the faucet nearest the frozen pipe so it can drain as it thaws. Waterproof the area with dropcloths in case leaks occur, then use one of the following methods to warm the frozen pipe (see **Drawing 97**). Be sure to work from the faucet toward the iced-up area.

- *Propane torch.* With a flame-spreading nozzle, the torch will quickly thaw a frozen pipe.
  CAUTION: Never let the pipe get too hot to touch.
- *Hair dryer.* Used like the torch, a dryer will gently defrost the pipe.
- *Heating pad.* A gradual but effective method, a heating pad wraps a length of pipe with warmth.
- *Heat lamp.* For freezes concealed behind walls, floors, or ceilings, beam a heat lamp 8 or more inches from the wall surface.
- *Hot water.* If no other method is available, wrap the pipe (except plastic) in rags and pour boiling water on it. Expect a mess.

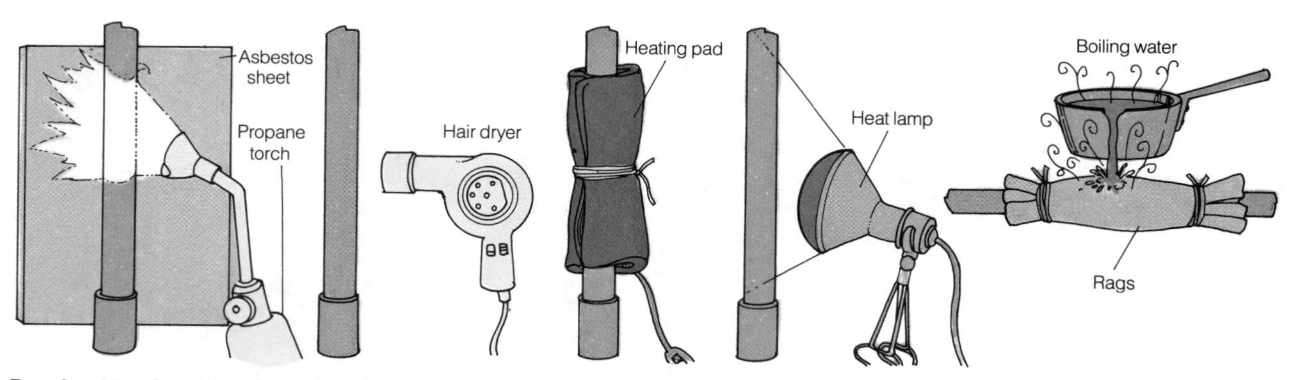

Drawing 97. A variety of pipe-thawing techniques

Closing down plumbing for the winter

Homeowners who used to simply turn down the thermostat in a vacated house for the winter are now closing down the plumbing system because of prohibitively high energy costs. Winterizing your plumbing is a virtually cost-free alternative to frozen pipes.

First, turn off the main shutoff valve (see page 23) or have the water company turn off service to the house. Starting at the top floor, open all faucets, indoor and out.

When the last of the water has dripped from the taps, open the plug at the main shutoff valve (if possible—you may have to contact the water company) and let it drain.

Turn off the power to the water heater and open its drain valve.

Empty water from the traps under sinks, tubs, and showers by opening cleanout plugs or removing the traps, if necessary (see pages 24–25). Empty toilet bowls and tanks, then pour a gallon of automotive antifreeze/water solution into each toilet bowl.

Finally, if your home has a basement floor drain or a main house trap, fill each with full-strength automotive antifreeze (see **Drawing A**).

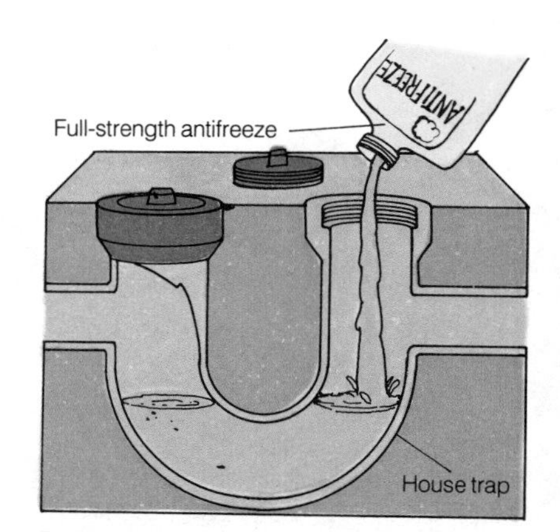

Drawing A. Fill the main house trap or basement floor drain with antifreeze.

Noisy Pipes

Pipe noises run up and down a nonmusical scale, ranging from loud banging to high-pitched squeaking, irritating chatter, and resonant hammering. Listen carefully to your pipes; the noise itself will tell you what measures to take to quiet the plumbing.

To get to the pipes, first look under the house. If you must cut into a wall or ceiling to reach a noisy pipe, refer to page 62.

Water hammer

The most common pipe noise—water hammer—occurs when you quickly turn off the water at a faucet or an appliance. The water flowing through the pipes simply slams to a stop, causing a shock wave and a hammering noise.

You can minimize or eliminate water hammer by installing air chambers—short, dead-end pieces of pipe—where there are none or by clearing existing chambers of water and residue.

Installing new air chambers. If the plumbing system in your house was designed without air chambers, or if some individual fixtures have none, buy them at a plumbing supply store and install them yourself. You can buy individual ones—known as air chambers, water hammer arresters, or air cushions—or purchase a master unit for the entire house.

Air chambers for individual fixtures are generally straight, capped lengths of pipe (see **Drawing 98A**) extending 2 feet up from supply pipes. Coiled air chambers (see **Drawing 98B**) for individual fixtures can be installed without tearing into walls.

Restoring air chambers. Most water systems have short sections of vertical pipe rising above each faucet or appliance. These sections hold air that cushions the shock when flowing water is stopped by a closing valve—the moving water rises in the pipe instead of banging to an abrupt stop. Sometimes these sections get completely filled with water and lose their effectiveness as cushions. To restore air chambers, take these steps:

- *Check the toilet tank* to be sure it is full; then close off the supply shutoff valve just below the tank.
- *Close the main shutoff valve* for the house water supply (see page 23).
- *Open the highest and lowest faucets* in the house to drain all water.
- *Close the two faucets;* reopen the main shutoff valve and the shutoff valve below the toilet tank. Normal water flow will reestablish itself for each faucet when you turn it on. (You can expect a few grumbles from the pipes before the first water arrives.)

Other pipe noises

Loud banging, chattering, or squeaking from your pipes indicates that wear or damage is taking place somewhere.

Banging. If you hear a banging noise whenever you turn on the water and it's not a water hammer problem, check the way the pipes are anchored. Banging pipes are generally easy to cure. You'll probably find the vibration-causing section of pipe is loose within its supports (see Drawing 23 on page 57).

To eliminate banging completely, slit a piece of old hose or cut a patch of rubber and insert it in the hanger or strap as a cushion (see **Drawing 99A**). For masonry walls, nail a block of wood to the wall with masonry nails, then anchor the pipe to it with a pipe strap (see **Drawing 99B**). It's a good idea to install enough hangers to support the entire pipe run (see pages 47–58); if there are too few supports, the pipe will slap against the flooring, studs, and/or joists.

Be careful not to anchor a pipe—especially a plastic one—too securely. Leave room for expansion with temperature changes.

Faucet chatter. This is the noise you hear when you partially open a faucet. To correct the problem, tighten or replace the seat washer on the bottom of the faucet stem (see page 14) to prevent the stem from vibrating in the seat.

Squeaking. Pipes that squeak are always hot water pipes. As the pipe expands, it moves in the support, and friction causes the squeak. To silence this, insulate between the pipe and supports with a piece of rubber as you would for banging.

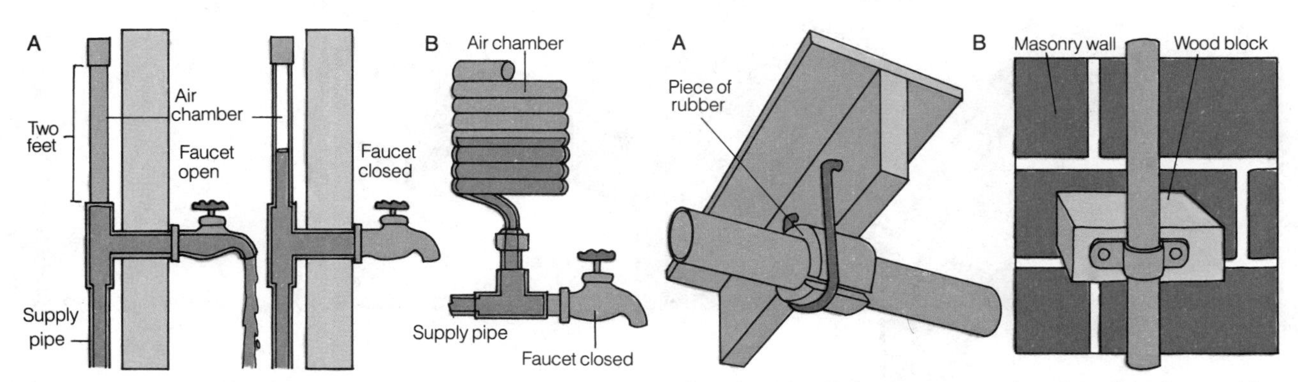

Drawing 98. Two types of air chambers: straight (A) and coiled (B).

Drawing 99. Two ways to stop pipes from banging: cushioning (A) and bracing (B).

Pipefitting Know-how

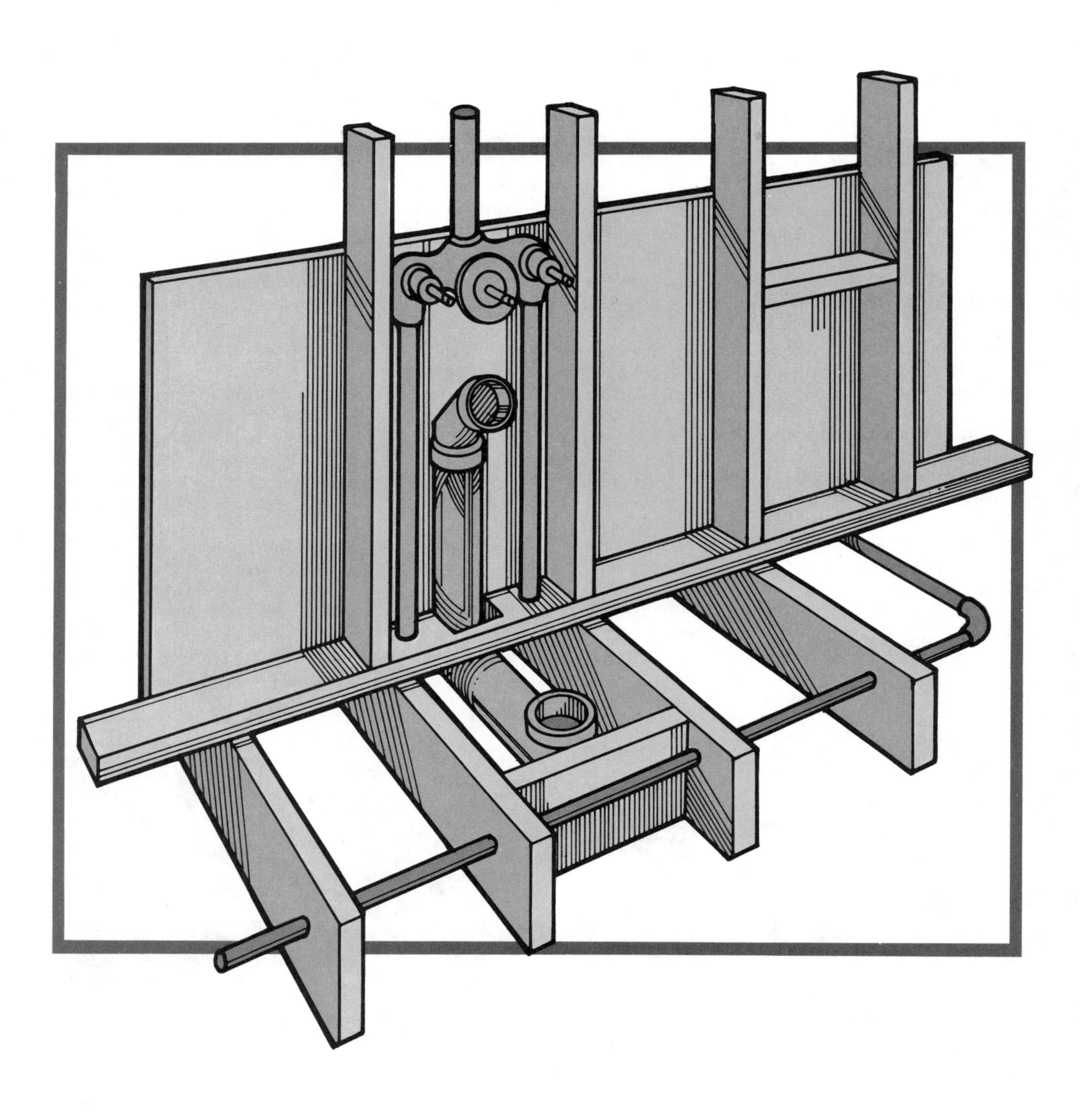

Working with Plastic Pipe

Ask any homeowner who has worked with various types of pipe what material he or she prefers, and the answer is likely to be "Plastic!" That's because plastic pipe is lightweight, inexpensive, and easy to cut and fit, making it ideal for do-it-yourself installations. It's also self-insulating and resistant to damage from chemicals and electrolytic corrosion—metal pipe is not. In addition, plastic's smooth interior surface provides less flow resistance than metal.

More and more local plumbing codes are allowing use of plastic pipe, but be sure to check with your plumbing inspector before installing it. Its use may be restricted because of possible health hazards, noise, and reaction to heat (some types of plastic go limp and change shape in contact with water over 140°.

Types of plastic pipe. Plastic pipe comes in both flexible and rigid types.

Flexible plastic types are PB (polybutylene) tubing, sometimes used for hot and cold water supply systems, and PE (polyethylene) tubing, sometimes used for cold water supply systems.

Rigid plastic pipe includes three common varieties: CPVC (chlorinated polyvinyl chloride), used for hot and sometimes cold water supply systems; PVC (polyvinyl chloride), used for cold water supply and DWV (drain-waste and vent) systems; and ABS (acrylonitrile-butadiene-styrene), used for DWV systems. A fourth variety, PP (polypropylene) pipe, currently is available for fixture traps and drainpipes only.

Pressure precautions. Plastic pipe has various pressure ratings—"schedules"—stamped right on the pipe; use the schedule number prescribed by your building inspector.

Because pressure ratings for plastic pipe are lower than those for metal pipe, plastic pipes are less able to withstand line surges (sudden changes in water pressure) in the water supply system. To prevent problems, install air chambers or air coils (see page 45) at all appropriate fixtures. Also, when using CPVC or PB pipe for hot water supply, replace the pressure and temperature relief valve (see page 38) on your water heater to match temperature and pressure ratings for these materials.

Removing plastic pipe. Since the use of plastic pipe indoors is a recent trend, chances are you'll encounter not plastic but copper pipe (see pages 52–55) or galvanized pipe (see pages 56–57) when you penetrate a wall (see page 62). If you do discover a run of plastic, it is likely to be the rigid type, which uses permanently glued fittings. This will require you to saw through the pipe itself.

Before cutting, have a bucket or absorbent cloth in place to catch any spills. Use a plastic pipe cutter or a backsaw, hacksaw or mini-hacksaw with a fine-toothed blade (24 or 32 teeth per inch). While you saw, brace the pipe to prevent excess motion that would strain joints; avoid letting the pipes sag when cutting is finished. After each cut, keep the cut ends dry by plugging them with a compressed lump of bread (once water flows through the pipes, the bread will dissolve).

Flexible supply pipe

Flexible plastic pipe is especially useful in cramped places because it can follow a winding course without requiring a lot of fittings.

The two types of flexible supply pipe used in homes—PB (polybutylene) for hot and cold lines of pipe, and PE (polyethylene) for cold lines only—are sold in rolls of 25 and 100 feet, or can be cut to the length needed. Though PB is more versatile and less expensive than PE, it may not be as readily available.

You can cut flexible plastic pipe with a sharp knife or an inexpensive special ratchet cutter. Then use one of several types of fittings to join it to a length of the same plastic pipe or another material.

Support horizontal runs of flexible plastic supply pipe every 6 to 8 feet, vertical runs every 8 to 10 feet. Your plumbing supplier will carry hangers that are made to support flexible pipe.

If you have to cut into existing flexible plastic pipe to replace or extend it, a sharp knife will give you the cleanest cut. Don't use a saw as you would with rigid plastic pipe.

CAUTION: Before beginning work on supply pipe, turn off the house water supply at the main shutoff valve (see page 23). Drain the pipes by opening a faucet at the low end.

Nipple fittings have hollow, corrugated nipples that fit into the ends of pieces of flexible pipe (see **Drawing 1**) and are held in place by stainless steel clamps.

Transition nipple fittings have a socket on one end.

Flexible pipe needn't be cut off squarely before nipple fittings are inserted.

If you need to make changes after the fittings are attached, usually the pieces can be pulled apart by hand once the clamps are loosened. If not, simply pour hot water over the ends of the pipe.

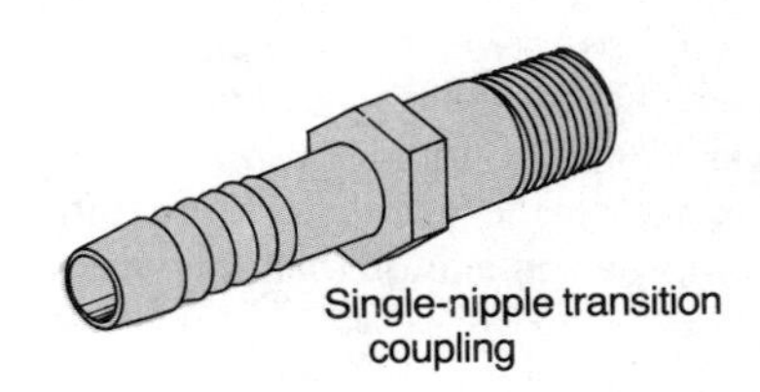

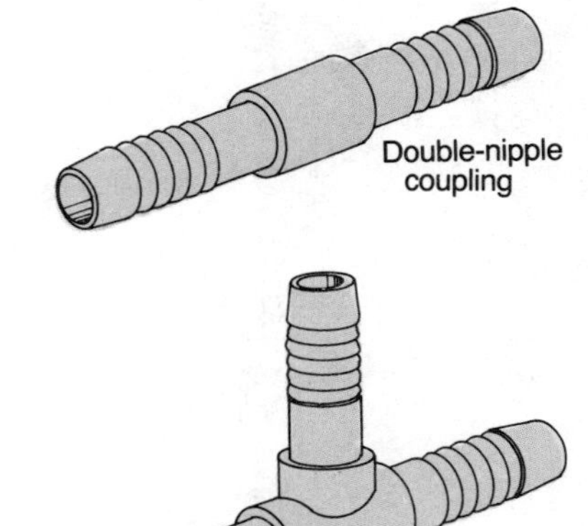

Drawing 1. Nipple fittings

. . . Plastic Pipe

Flare fittings of the type used with copper pipe (see page 53) can also be used with flexible plastic, since both types of pipe come in the same sizes. Use them as transition fittings —to join plastic to copper—or simply to join flexible plastic pipes to each other. You may be able to find flare fittings made of plastic (less expensive than the more common bronze fittings). Flare fittings are especially useful for connecting pipe to faucets and appliances.

To make flare fittings watertight, pipe must be cut off squarely (with a plastic pipe cutter or in a miter box, using a hacksaw, mini-hacksaw, or backsaw with a fine-toothed blade—24 or 32 teeth per inch). After slipping on the flare nut, you warm the end of the pipe in water, then flare it with a standard copper-pipe flaring tool (see Drawing 15 on page 55). When the flared end has cooled, slip the push-on socket or screw the threaded nipple (see **Drawing 2**) onto the pipe to be joined; then screw the flare nut onto the nipple. Use two crescent wrenches to tighten.

Compression fittings, often used with copper pipe (see page 55), can also be used with flexible plastic. Use them as transition fittings—to join plastic to copper—or simply to join flexible plastic pipes to each other. Compression fittings (see **Drawing 3**) have more components than flare fittings but are simpler to assemble because the end of the flexible plastic pipe doesn't need flaring. Though they're commonly made of bronze, compression fittings are sometimes available in plastic. Compression fittings are especially useful for connecting pipe to faucets and appliances. Before assembling, be sure to cut the end of the pipe off squarely (with a plastic pipe cutter or in a miter box, using a hacksaw, mini-hacksaw, or backsaw with a fine-toothed blade—24 or 32 teeth per inch).

To assemble a compression fitting, slip the connector nut over the end of a length of pipe, so that the nut's broad shoulder will be closest to the end. After the nut, slip on the compression ring, flange, and O-ring. With your fingers, screw the connector nut onto the threaded cone of the corrugated nipple; tighten with a crescent wrench on the connector nut and a pipe wrench on the threaded cone.

One-step connectors, molded of acetate and containing internal stainless steel gripper rings, require no tools, clamps, or solvent-cement. You simply push the pipe into the fitting until it locks into place. One-step connectors resemble the cemented push-on fittings used on rigid supply pipe (see **Drawing 4**, on facing page). More expensive than nipple fittings, but less expensive than flare and compression fittings, they are available as couplings, elbows, and tees in a variety of sizes.

Though they were developed for use with flexible plastic pipe, one-step connectors work equally well on rigid plastic pipe (following) and on hard and soft copper pipe (page 52). Before purchasing, make sure these connectors meet the requirements of your local plumbing code.

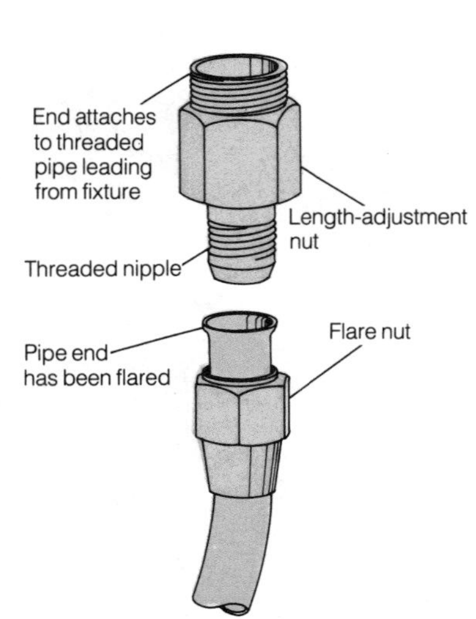

Drawing 2. Flare fitting

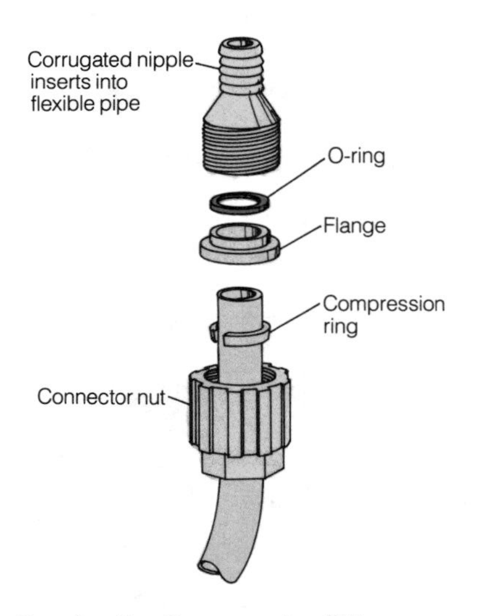

Drawing 3. Compression fitting

Rigid supply pipe

The two types of rigid plastic supply pipe used indoors are PVC (to transport cold water only) and CPVC (to transport hot and cold water). Both types are somewhat limber and can follow slight directional changes without cracking.

Available in 10 and 20-foot lengths, rigid pipe can be bought by the piece or in bundles of ten. Common sizes are ½-inch, ¾-inch, and 1-inch nominal diameters (the diameter used in matching fittings, even though actual diameter may be different). Look for the stamp of approval by the National Sanitation Foundation on any rigid supply pipe bought for indoor use.

Take care not to store the pipe in direct sunlight for longer than a week, because the accumulation of ultraviolet rays can make it brittle.

If you have to cut into existing runs of rigid plastic supply pipe to replace or extend it, use the techniques described under "Removing plastic pipe," page 47.

CAUTION: Before beginning work on a supply pipe, turn off the house water supply at the main shutoff valve (see page 23). Drain the pipes by opening a faucet at the low end.

Fittings. Most PVC and CPVC fittings (see **Drawing 4**) push onto the ends of the pipes and are cemented in place with a permanent solvent-cement. Transition fittings—those that let you link plastic pipe to pipe of a different material—often have threads on one end. Threaded fittings called unions make it easy for you to change or extend supply pipe by unscrewing a length of pipe in the middle of a run. Reducer fittings allow you to link pipes of different diameters.

Measuring, cutting, and hanging pipe. Before you cut any pipe, be sure your measurements are exact. Rigid pipes won't "give" to compensate if they're too long, and no one has invented a pipe stretcher to take care of too-short lengths.

You need to determine the face-to-face distance between new fittings, then add the distance the pipe will extend into the fittings (see **Drawing 5**). If you use two short pieces of pipe connected by a union, don't forget to count the union as a third fitting.

The distance the new pipe will extend into the fittings depends on the nature of the fittings. In push-on fittings, pipe ends extend all the way to the beginning of the shoulder once the solvent-cement is brushed on; in threaded fittings, they don't go quite so far. You can most easily cut rigid plastic pipe with a plastic pipe cutter or in a miter box, using a hacksaw, mini-hacksaw, or backsaw with a fine-toothed blade—24 or 32 teeth per inch (see **Drawing 6**).

Once installed, horizontal runs of rigid plastic supply pipe should be supported every 6 to 8 feet with one of the hangers shown in Drawing 23, page 57. Because of the pipe's light weight and rigidity, vertical runs require no support.

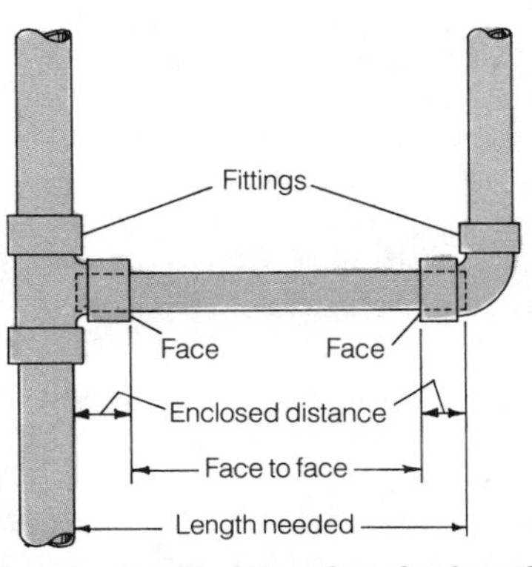

Drawing 5. To determine pipe length, add face-to-face to enclosed distance.

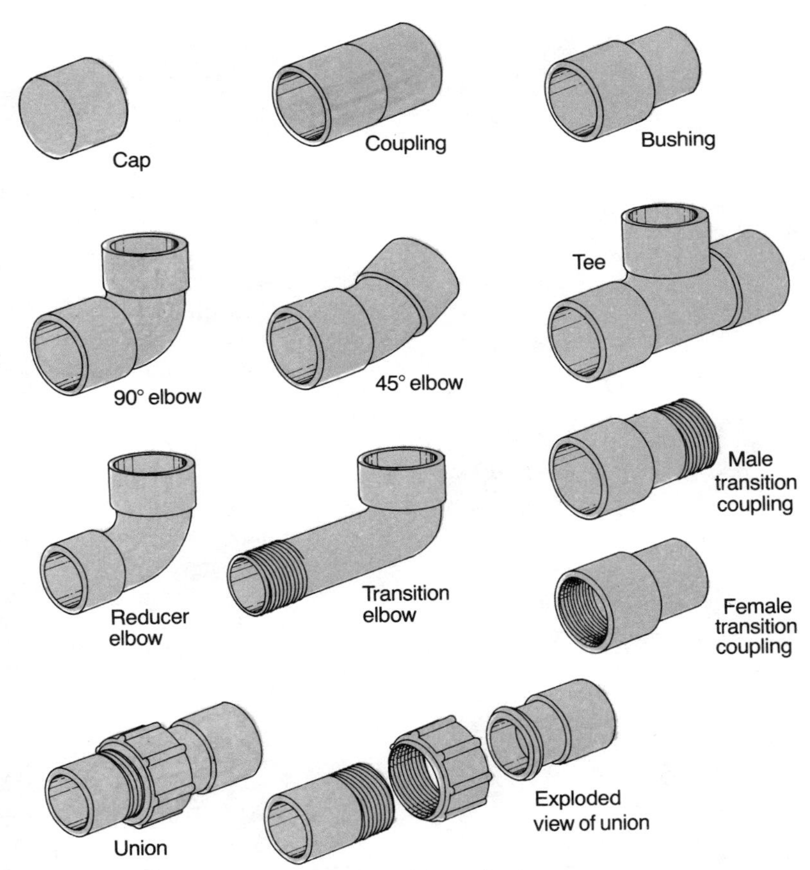

Drawing 4. PVC and CPVC fittings for rigid supply pipe

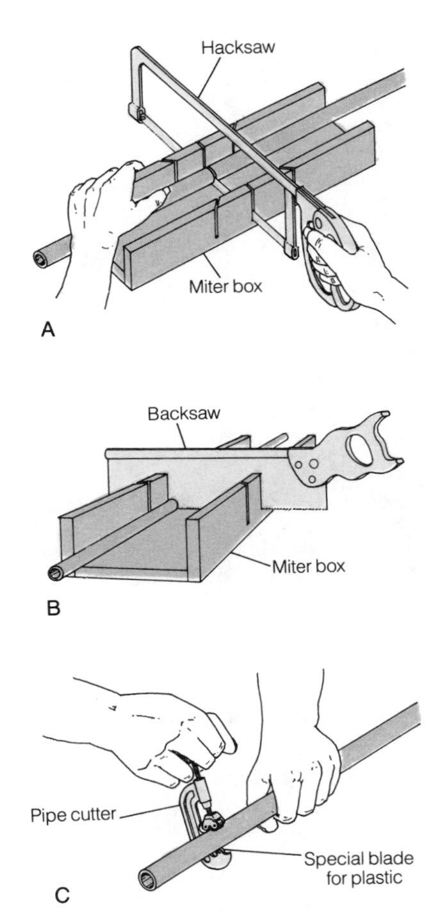

Drawing 6. To cut plastic pipe squarely, use a hacksaw and miter box (A), a backsaw and miter box (B), or a pipe cutter (C).

. . . Plastic Pipe

Cementing push-on fittings. First remove all burrs from the cut end with a knife or piece of sandpaper (see **Drawing 7A**); clean the end with a rag. Use sandpaper (or apply primer manufactured for the purpose) to remove the gloss from the last inch of the cut end's outside surface (see **Drawing 7B**). Slightly roughening the surface will help the joint hold.

Be sure to get the right type of solvent-cement for the kind of plastic you're using. Usually the container will have an applicator brush fastened to its top. (If a brush isn't included, use any soft brush—a ½-inch brush for ½-inch pipe, ¾-inch brush for ¾-inch pipe, and so forth.) When working, be sure to keep lighted matches and cigarettes away from the flammable solvent-cement, and avoid breathing fumes.

Before cementing the pipe and fitting together, you should know exactly how the finished run will line up (solvent-cement hardens very fast). It's a good idea to mark the pipe and the fitting beforehand and line up the two marks (see **Drawing 7C**) when cementing. The pipe won't slide into the fitting completely until the cement (which acts as a lubricant) is applied; make marks long enough to take this into account.

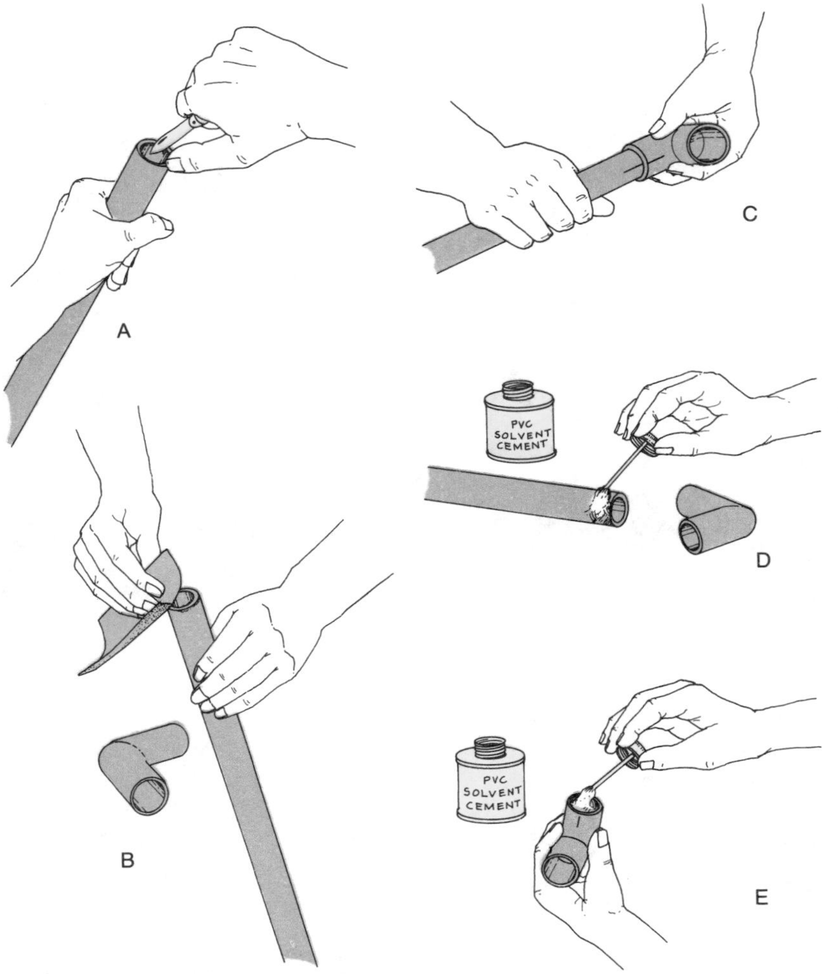

Drawing 7. To cement push-on fittings, use knife to pare away burrs (A); sand gloss from last inch (B); place fitting over end and mark for placement (C); apply solvent-cement to outside of pipe (D); then coat inside of fitting (E).

Apply a heavy coat of solvent-cement to the last inch of the pipe's outside surface (see **Drawing 7D**); brush a light coat on the inside of the fitting (see **Drawing 7E**). Put the two pieces together immediately. Give them a quarter-turn to spread the cement evenly; then line up the marks precisely. The bead of excess cement should be uniform all around the lip of the fitting. If there's no bead or an uneven bead, the joint needs more solvent-cement. If there's a very heavy bead, there's probably too much solvent-cement in the fitting; wipe the excess cement from around the lip of the fitting.

Hold the fitting and pipe together for about a minute. Then, if possible, check inside the fitting with your finger for excess solvent-cement, and use a knife to cut away any excess. Wait at least an hour before you put any water into the pipe—if air temperature is between 20° and 40°F/−6° and 4°C, wait at least 2 hours; between 0° and 20°F/−18° and −6°C, wait at least 4 hours.

Attaching screw-on fittings. Once in a while you'll encounter rigid plastic pipe whose ends have exterior threads (most common with the 1 to 12-inch precut lengths of pipe known as nipples). Such pipe requires special plastic fittings with interior threads. To make a watertight seal with screw-on fittings, first wrap pipe-wrap tape one and a half turns clockwise around the threads of the pipe, pulling the tape so tight that the threads show through, before screwing on the fitting. (Do not attempt to coat the interior threads of the fitting.)

Rigid drain-waste and vent (DWV) pipe

Two types of plastic pipe are used to repair or extend DWV systems—PVC (polyvinyl chloride), and ABS (acrylonitrile-butadiene-styrene). You can tell the difference by their color: PVC is off-white, ABS is black. (Basin traps and other short connectors to DWV lines are often fabricated from a third plastic, polypropylene—PP—

identified by its stark white color.) Both PVC and ABS are less expensive, lighter in weight, and easier to connect and hang than sections of cast iron pipe (see page 58). For these reasons, plastic pipe is a common choice for extending a cast iron system, and even for replacing a leaking cast iron pipe.

If you can, choose PVC over ABS pipe. PVC is less susceptible to mechanical and chemical damage, will not burn, and has a slightly greater variety of fittings available (the fittings for PVC and ABS are not interchangeable).

Sold in lengths of 5, 10, and 20 feet, plastic drainpipe for tubs, sinks, and lavatories normally has a 1½ or 2-inch nominal diameter, for toilets and house drains a 3 or 4-inch nominal diameter. Vent pipe can range from 1¼ to 4 inches in nominal diameter, depending on how many pipes penetrate the roof.

Plastic DWV pipe should be supported at every fitting or every 4 feet, whichever is less; use the plumber's tape shown in Drawing 23, page 57, or a jumbo version of the wire hanger shown in the same drawing.

If you need to remove a section of cast iron DWV pipe to make room for plastic pipe, see page 58. To remove a section of plastic DWV pipe, see "Removing plastic pipe," page 47. Measure and cut plastic DWV pipe in the same way you measure and cut plastic supply pipe (see "Measuring, cutting, and hanging pipe," page 49). Join plastic DWV pipe in the same way you join plastic supply pipe (see "Cementing push-on fittings," on the facing page). Remember to allow ¼ inch of downward slope toward the house drain for every foot of DWV pipe run horizontally (see page 62–63).

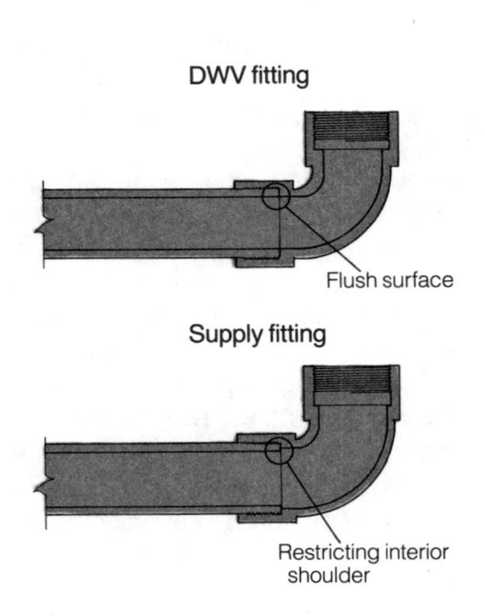

Drawing 8. DWV fitting has no shoulder to block smooth flow of waste water; supply fitting has a shoulder.

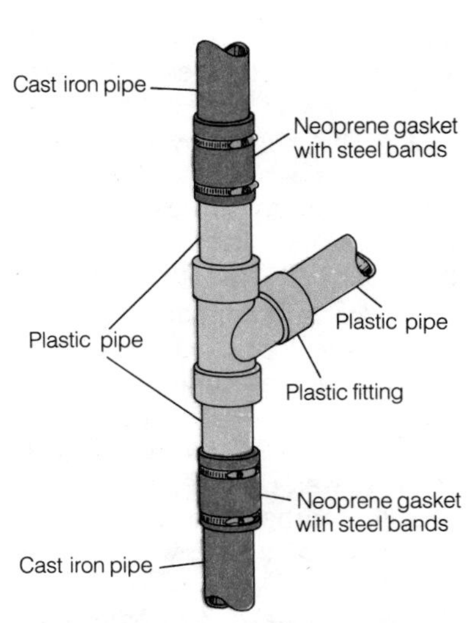

Drawing 9. To join plastic pipe to cast iron, connect as shown to short lengths of plastic pipe.

DWV fittings (also called sanitary fittings) differ from supply fittings in that they have no interior shoulder that could catch waste (see **Drawing 8**). You use a different method to assemble DWV fittings, depending on whether you're joining plastic to plastic, plastic to lengths of hubless cast iron, or plastic to the bell of bell and spigot (hub) cast iron. Fittings for joining plastic to plastic and plastic to lengths of hubless cast iron (see **Drawing 9**) include reducer fittings that allow you to connect pipes of different diameters; always take care to direct the flow from the lesser to the greater pipe diameter.

Joining plastic to plastic. As with plastic supply pipe, solvent-cement is used to connect the pipe to its fittings (see "Cementing push-on fittings," facing page). If you're adding DWV pipe to your present system, make sure the cut pipe already in the wall or floor is braced securely before you add new pieces. To brace the pipe, use the plumber's tape shown in Drawing 23 on page 57, or a jumbo version of the wire hanger shown in the same drawing.

There may not be enough clearance or slack between the free end of a new fitting and the pipe already installed to slip the fitting over the pipe. In this case, you must cut the pipe back and make the connection either with spacers (stubs of plastic pipe) and slip fittings (see page 63), or by installing a union.

Joining plastic to hubless cast iron. Cut out the section for replacement or a new fitting as described on page 58. To replace a damaged cast iron run with plastic, cut the plastic pipe to the exact length of the opening, place the new run in position (you may need a helping hand), and, if codes permit, make the connections with neoprene gaskets and stainless steel bands (see "Putting cast iron pipe together," page 58).

To extend new plastic pipe from a cast iron system, you have a choice of methods. You can add a hubless sanitary tee with gaskets and bands (Drawing 32A, page 63) and run plastic from that point; or, if codes permit, you can install a plastic sanitary tee with spacers (stubs of plastic pipe), gaskets, and bands (see **Drawing 9**).

Joining plastic to the bell of bell and spigot cast iron. To remove a length of cast iron pipe from a bell, cut out a piece of the cast iron (see page 58), leaving a short length in the bell. Remove the short length by working it back and forth while you heat the joint with a blow torch to melt the lead. To join the plastic to the bell, either pack the joint with oakum and then hammer in cold lead wool to make the joint waterproof, or pack a puttylike plastic substitute for oakum and lead into the joint with a putty knife. These materials are available at plumbing supply houses.

Working with Copper Pipe

Copper pipe, often called copper tubing, is lightweight, fairly easy to join (by soldering or with flare, compression, or union fittings), highly resistant to corrosion, and rugged. Its smooth interior surface minimizes resistance to water flow and to small particles that may occasionally slip into the pipes.

Two kinds of copper pipe—hard and soft—are used in supply systems to carry fresh water. Another kind—with larger diameter—is used in drain-waste and vent (DWV) systems to carry used water, waste, and noxious air. Still another type of copper pipe—corrugated supply—is used as flexible tubing to link hard or soft pipe to fixtures. However, in some areas this type of pipe is difficult to find.

- ***Hard supply pipe*** is sold in lengths of 20 feet or less. Because it can't be bent without crimping, it must be cut and joined with fittings whenever a change in direction is made. It comes in three thicknesses: K (thick wall), L (medium wall), and M (thin wall); M is usually adequate for above-ground plumbing. Nominal diameters range from ¼ to 1 inch; actual diameters may be more.
- ***Soft supply pipe*** is sold in 60-foot coils. More expensive than hard supply pipe, it offers a big advantage: it can be bent around curves without crimping, and thus without the use of fittings. You can buy it in two thicknesses: K (thick wall) and L (medium wall). L is usually adequate for above-ground plumbing. Nominal diameters range from ¼ to 1 inch; actual diameters may be slightly more.
- ***Copper DWV pipe*** is usually sold in 20-foot lengths and in nominal diameters of 1½ and 2 inches (copper DWV pipe with nominal diameters larger than 2 inches is too expensive to be readily available in most geographical areas).
- ***Flexible tubing*** of corrugated or smooth copper or chrome-plated copper comes in short lengths for linking supply pipe to fixtures. Able to conform to tighter curves than soft copper pipe, flexible tubing has a nominal diameter of ⅜ or ½ inch. It often comes in kit form; follow the manufacturer's instructions. With some designs, you slip the special nut that comes with the fixture over the stub end of pipe leading from the fixture, then place the flared fitting and rubber washer attached to the tubing against the end of pipe leading from the fixture. You seal the joint by tightening the nut with an adjustable wrench.

Removing, measuring & cutting pipe

To replace leaking copper pipe, or to extend supply pipes, you'll need to learn some basic techniques.

The following techniques of removing, measuring, and cutting copper pipe apply equally to hard and soft supply pipe. Drain-waste and vent (DWV) pipe is measured and cut in the same ways, but it usually needs to be braced like cast iron DWV pipe (see "Drain-waste and vent connections," page 63) before removal.

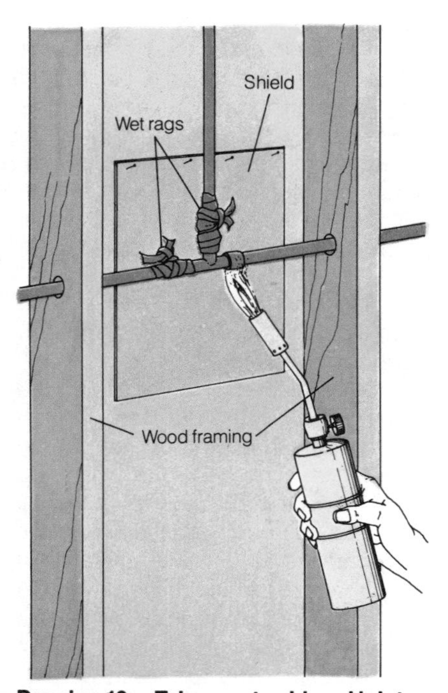

Drawing 10. Take apart soldered joints with a torch if pipe's position will allow ends to be pulled free.

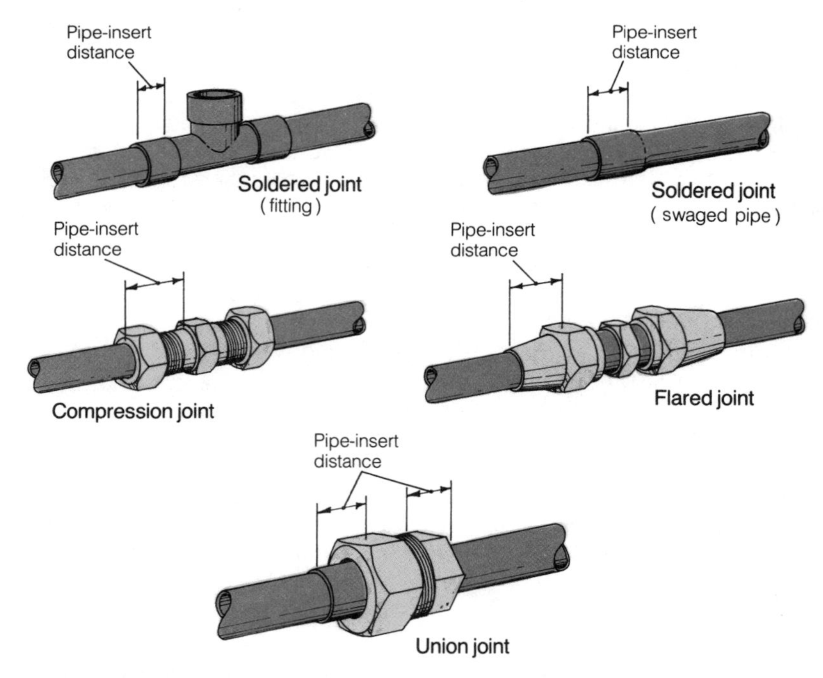

Drawing 11. Pipe-insert distances for various types of copper joints. Be sure to add insert distance to total distance between new fittings to determine length of pipe.

CAUTION: Before beginning work on supply pipe, turn off the house water supply at the main shutoff valve (see page 23). Drain the pipes by opening a faucet at the low end.

Removing copper pipe. Make sure the pipe run is braced, both to prevent excess motion that would strain joints and to keep severed ends from sagging. It's often possible to achieve sufficient bracing by wrapping lengths of plumber's tape (as shown in Drawing 23 on page 57) once around the circumference of pipe every 3 feet, then pulling the tape taut and nailing it to nearby joists or studs.

The fastest way to disassemble a run of soldered copper pipe is to make a straight cut through the pipe with a fine-toothed (24 or 32 teeth per inch) hacksaw. But if the pipe is in a position where the ends can be pulled free once the solder joint is loosened, you can melt the solder with a butane or propane torch (soft pipe is usually easy to pull free). Before applying the torch, be sure to shield flammable material with a piece of metal and wrap wet rags around joints you wish to leave intact (see **Drawing 10**). In tight quarters, soldering can be difficult.

Compression and flare fittings, as well as unions, uncouple by unscrewing.

Measuring copper pipe. To determine how much new copper pipe you need, measure the distance between new fittings, then add the distance the pipe will extend into the fittings. Pipe-insert distances vary for various types of joints (see **Drawing 11**).

Cutting copper pipe. Plan to cut new lengths of copper pipe with a pipe cutter (see Drawing 6C on page 49) that has a blade designed for copper pipe. To use the cutter, twist the knob until the cutter wheel just pierces the copper surface. Rotate the cutter around the pipe, tightening after each revolution, until the pipe snaps in two.

You can also cut copper pipe with a fine-toothed (24 or 32 teeth per inch) hacksaw or mini-hacksaw, but it's more difficult to make a straight cut with a saw than with a pipe cutter.

After cutting the pipe, clean off inside burrs with a half-round file or with the retractable reamer sometimes found inside pipe cutters. File or sand off outside burrs.

Joining & hanging copper pipe

Whether the pipe is "hard" or "soft," all copper is a soft metal, so you'll want to be careful not to damage it as you work. Use electrician's tape to cover vises that have teeth and the jaws of wrenches.

Different joining methods require different fittings (see **Drawing 12**). Soldering—using the necessary fitting or a swaged joint—is the best way to keep copper pipe (both hard and soft) together, but in tight quarters soldering is sometimes difficult. Hard supply pipe can be joined with compression fittings, soft supply pipe with compression or flare fittings. Copper's softness prevents it from being successfully threaded.

If you think you'll need to remove a run of copper pipe regularly—to replace a gate valve, for instance (see page 22)—you'll want to fit two short lengths of the copper pipe together with a union (information about union joints is on page 55).

Reducer fittings allow you to link pipes of different diameters; transition fittings let you link copper pipe with plastic or galvanized pipe. (If you link copper with galvanized, though, you must use special fittings called dielectric fittings, to prevent electrolytic corrosion. See Drawing 19 on page 56.)

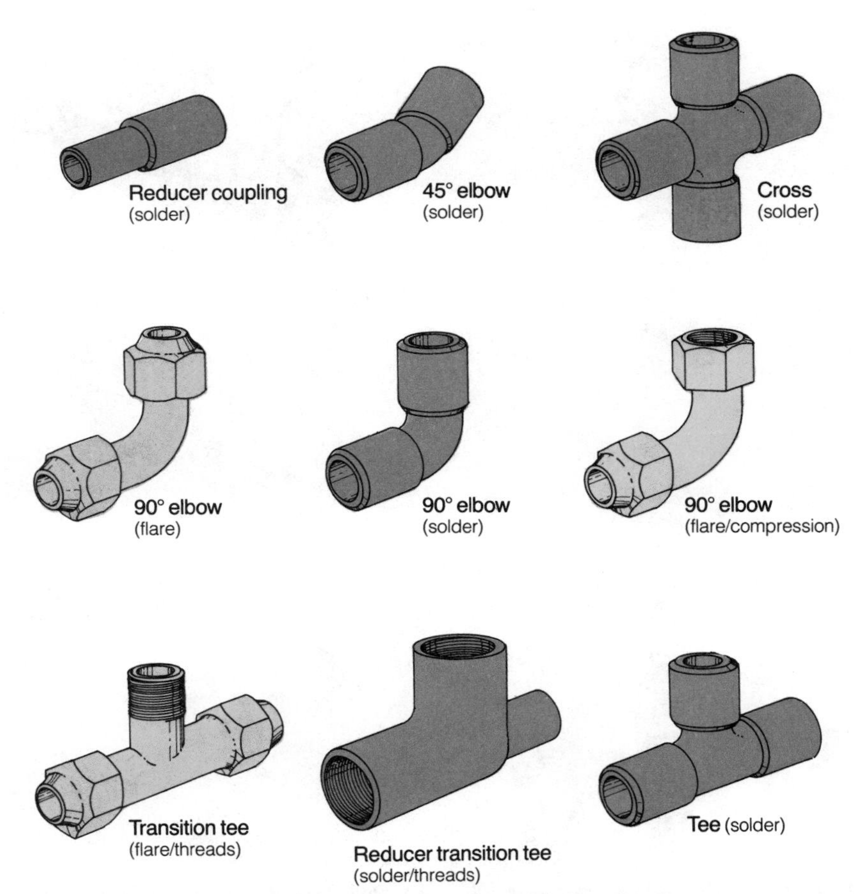

Drawing 12. Copper fittings. Transition tees are threaded when they connect copper pipe to galvanized pipe.

. . . Copper Pipe

Soldered joints. Often called sweat joints, soldered joints can be made with copper fittings that have smooth interiors, or by expanding one end of soft pipe to receive another—a process known as swaging (see **Drawing 13**).

To swage copper pipe, slip the swaging tool into one end of the pipe; then, with a ball peen hammer (shown in Drawing 13) or a soft-headed steel hammer, hammer the tool into the pipe to the point where the head of the tool reaches its greatest diameter (ends its outward flare).

NOTE: For simplicity's sake in what follows, we use the term "fitting" to refer equally to an actual fitting on a length of hard or soft copper pipe and to the expanded end of swaged soft copper pipe.

To solder a joint, you'll need a small butane or propane torch, some 00 steel wool or very fine sandpaper or emery cloth, a can of soldering flux, and some solid-core wire solder. CAUTION: Use only solder that's labeled lead-free in making joints in any potable (drinking) water supply system.

If there's any water in the pipes, it will hinder a successful soldering job. Dry the pipes as much as possible by turning off the water supply at the main shutoff valve (see page 23), then opening a faucet at the low end of the pipes. Stuff the ends of the pipe with plain white bread to absorb any remaining moisture. The bread will disintegrate once the water is turned on.

Use the steel wool, sandpaper, or emery cloth to polish the last inch of the outside end of the pipe and the inside end of the fitting down to the shoulder until they are shiny (see **Drawing 14A**). With a small, stiff brush, apply flux around the polished inside of the fitting and around the polished outside of the pipe end (see **Drawing 14B**).

Place the fitting on the end of the pipe. Turn the pipe or the fitting back and forth once or twice to spread the flux evenly. Next, position the fitting correctly and heat it with a torch, moving the flame back and forth across the fitting to distribute the heat evenly (see **Drawing 14C**). It's important not to get the fitting too hot because the flux will burn—simply vanish—if it's over-

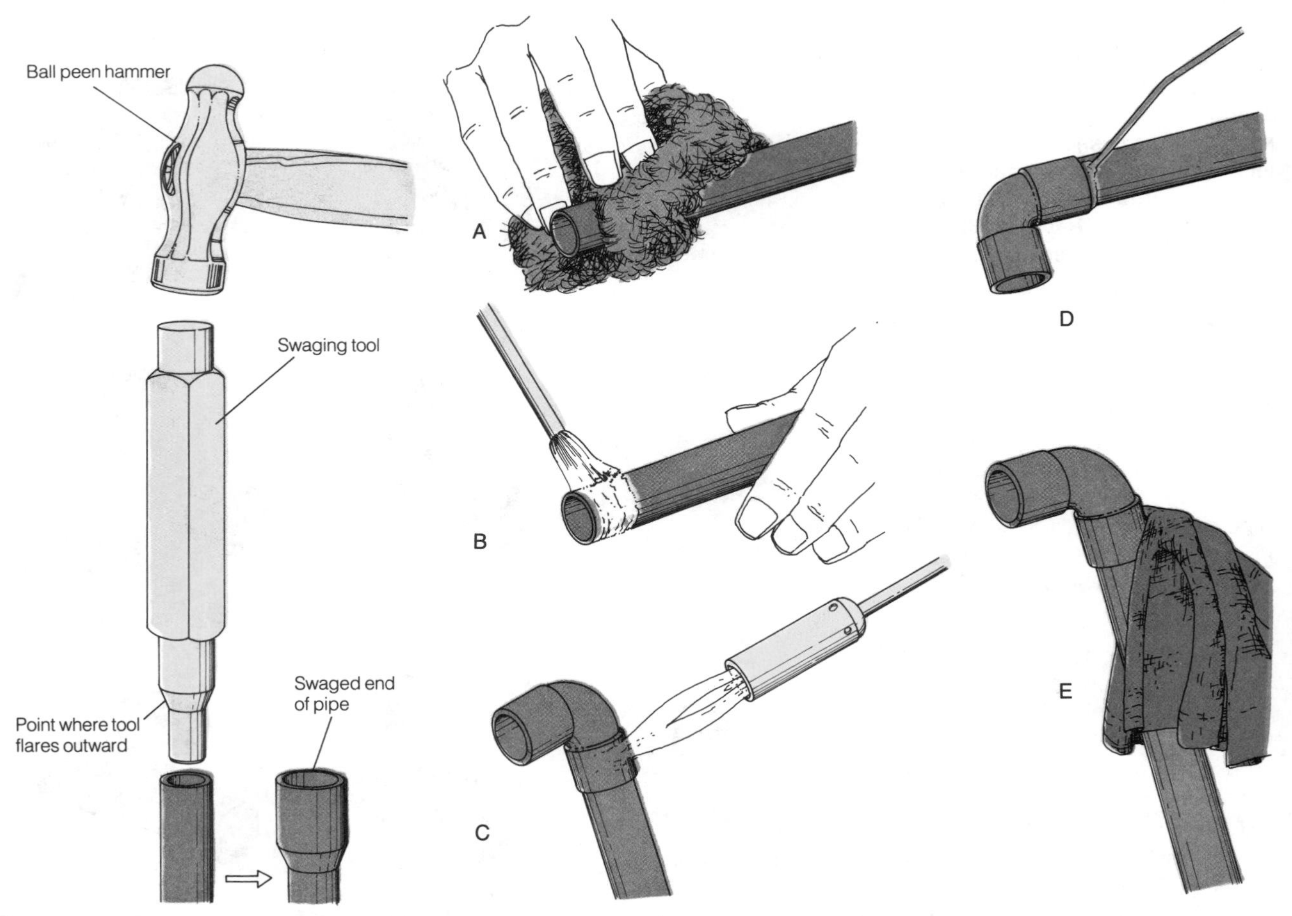

Drawing 13. To swage copper pipe, slip swaging tool into one end, then hammer it in to the point where tool ends its flare.

Drawing 14. To solder copper pipe, first polish outside end of pipe with steel wool (A). Apply flux (B); place fitting on end of pipe and heat (C). Touch solder to edge of fitting (D); wipe off excess with rag (E).

heated. Test the heat level this way: the joint is hot enough when solder will melt on contact with it. Touch the wire of solder to the joint occasionally as you're heating. The instant the wire melts, the joint is ready.

Take the torch away and touch the solder to the edge of the fitting; capillary action pulls solder in between the fitting and the pipe (see **Drawing 14D**). Keep soldering until a line of molten solder shows all the way around the fitting. With a damp rag, wipe off the surplus solder before it solidifies, leaving just a trace of solder showing in the crevice between fitting and pipe (see **Drawing 14E**). Keep your hands well away from the joint—the pipe gets quite hot within a foot or two of the joint. Be careful not to bump or move the newly soldered joint.

Flared joints. A flared joint works only on soft copper pipe. Since it tends to weaken the end of the pipe, use a flared joint only if you can't solder and can't find the right compression fitting.

To make a flared joint, slide the flare nut over the end of the pipe, tapered end facing away from the end of the pipe (see **Drawing 15A**). Clamp the end of the pipe into a flaring tool and screw the ram into the end of the pipe (see **Drawing 15B**). Remove the pipe from the flaring tool. Press the tapered end of the body of the fitting into the flared end of the pipe, and screw the nut onto the body of the fitting (see **Drawing 15C**). Use two wrenches to tighten.

Compression joints. The advantage of using compression fittings instead of flared fittings to create nonsoldered joints is that compression fittings work equally well on hard and soft copper pipe. A further advantage is that, unlike flare fittings, compression fittings don't require a special tool to assemble.

To install a compression fitting, slide the compression nut over the end of the pipe, broad shoulder facing away from the end. Then slip on the compression ring (see **Drawing 16**). Push the threaded body of the fitting against the end of the pipe, and screw the nut onto the body of the fitting. Tighten the nut with two wrenches, one on the nut and one on the body of the fitting. This compresses the ring tightly on the end of the pipe, effecting a watertight seal. Once compressed, the ring can't be removed.

Union joints. Like compression fittings, unions are composed of three elements. Unlike compression fittings, the three elements of union joints allow you to install or remove a union without having to twist the pipe itself. This makes a union especially easy to remove. Unions link pipes of the same diameter only and are available only as straight couplings.

To assemble a union, solder the male shoulder onto one length of pipe, then slip the nut onto the second length. Solder the female shoulder onto the end of the second length of pipe. Bring the male and female shoulders together, then slide the nut over the female shoulder, and screw the nut onto the male shoulder (see **Drawing 17**). To tighten, use two wrenches—one holding the pipe, the other turning the nut.

Hanging the pipe. Once lengths of pipe are joined, support the run every 6 to 8 feet with one of the hangers shown in Drawing 23 on page 57. Be sure first to insulate the pipe from the hanger so that electrolysis can't occur; do this by wrapping the pipe with electrician's tape where the hanger would touch it.

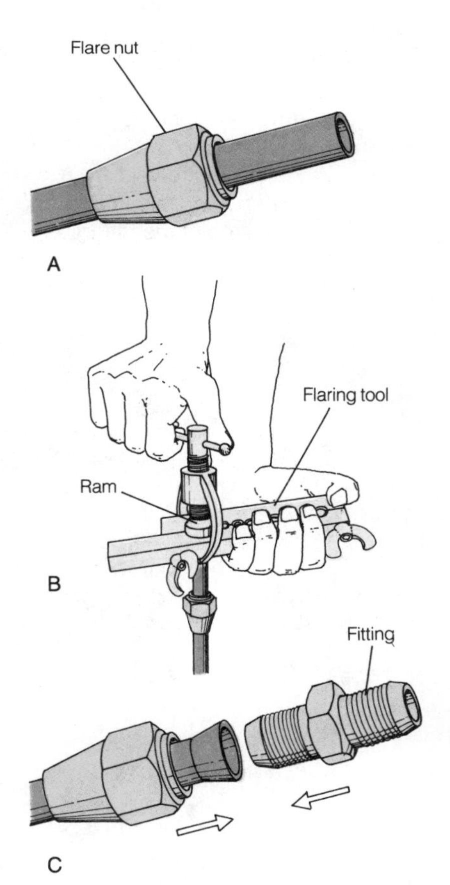

Drawing 15. To make flared joint, slide nut over pipe (A); clamp pipe into flaring tool, screw in ram (B). Screw on nut (C).

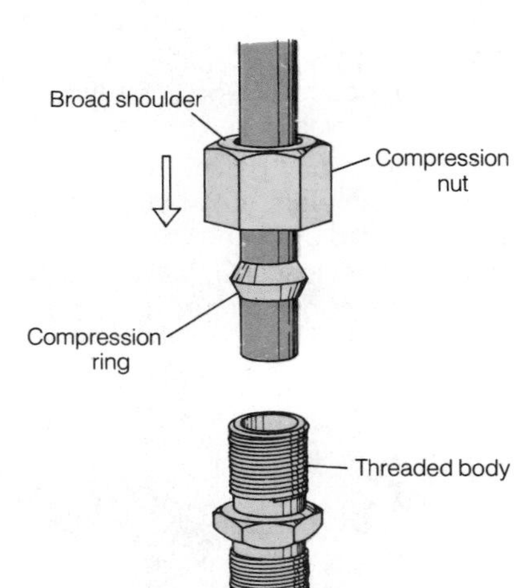

Drawing 16. To make compression joint, slide nut over pipe and slip on ring. Screw nut onto threaded body; tighten.

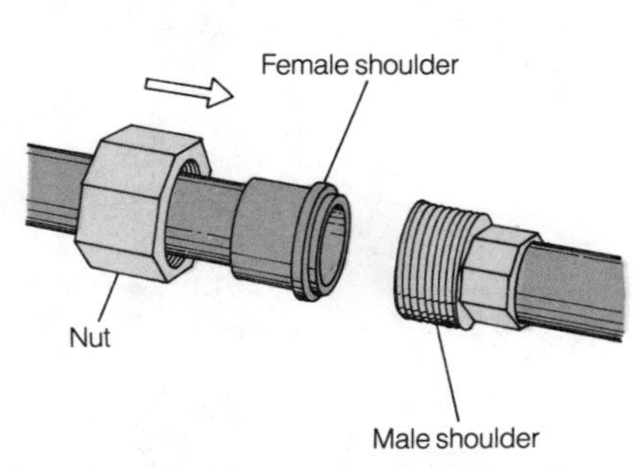

Drawing 17. To make union joint, slip nut over pipe and solder shoulders onto pipes. Join pipes with the nut.

Working with Galvanized Pipe

If your house is more than 20 years old, chances are you have galvanized supply pipe; you may even have galvanized drain-waste and vent pipe, though cast iron DWV pipe (see page 58) is more common.

The term "galvanized" means that iron pipe and fittings have been coated with zinc to resist corrosion. In spite of this, galvanized pipe not only corrodes faster than cast iron or copper, but because of its rough interior surface, it collects mineral deposits that can impede the flow of water.

It's standard procedure to replace a leaking length of galvanized pipe with the same type of pipe. To do this requires less equipment and expense than using copper or cast iron, and, unlike plastic pipe, galvanized pipe always conforms to local building codes.

But if you want to extend a supply system of galvanized pipe, use copper (see pages 52–55), or, if local code allows, plastic (see pages 47–51).

You can buy galvanized pipe in nominal diameters of ¼ to 2½ inches and in lengths of 10 and 21 feet, or cut to your specifications. Most retailers also carry short pieces of pipe—called nipples—in ½-inch increments from 1½ to 6 inches long, and then in 1-inch increments up to 12 inches long, and in diameters that match the pipes.

Threading. Galvanized pipe is connected to fittings by means of threads. Galvanized pipe is sold threaded on both ends. When you have pipe cut at the store, you might be able to have its new ends threaded there. If not, you can rent special tools for threading pipe (see facing page).

Using fittings. A great many fittings (see **Drawing 18**) are available for linking galvanized with galvanized, and galvanized with copper or plastic. If you cut into a galvanized pipe, you'll have to reconnect the ends with a union—a special fitting that lets you connect two threaded pipes without having to turn them. If you plan to add copper pipe to galvanized, use the coupling called a dielectric union (see **Drawing 19**). It contains an insulating washer and an insulating sleeve to keep electrolysis from occurring between the galvanized and the copper portions of the fitting.

Using wrenches. Working with galvanized pipe requires the simultaneous use of two pipe wrenches. If you're removing cut pipe, use one wrench to grasp the fitting, the other to grasp the pipe inserted into the fitting. Rotate only one of the wrenches. A fitting always screws on clockwise, off counter-clockwise. Be sure to apply force *toward*, rather than away from, the wrenches' jaws (see **Drawing 20**).

It's important to choose wrenches of the proper size. Disassembling old pipe requires the largest wrench you can find. Assembling pipe requires a 12 to 14-inch wrench for ½ to 1-inch pipe, an 18-inch wrench for 1¼ to 1½-inch pipe. Avoid wrenches with soft or dull teeth; they can cause muscle strain.

Removing, measuring & cutting pipe

Though all galvanized pipe is measured and cut in the same way, there are distinctions for removing it. On the next page we tell you how to remove supply galvanized pipe. You remove drain-waste and vent (DWV) gal-

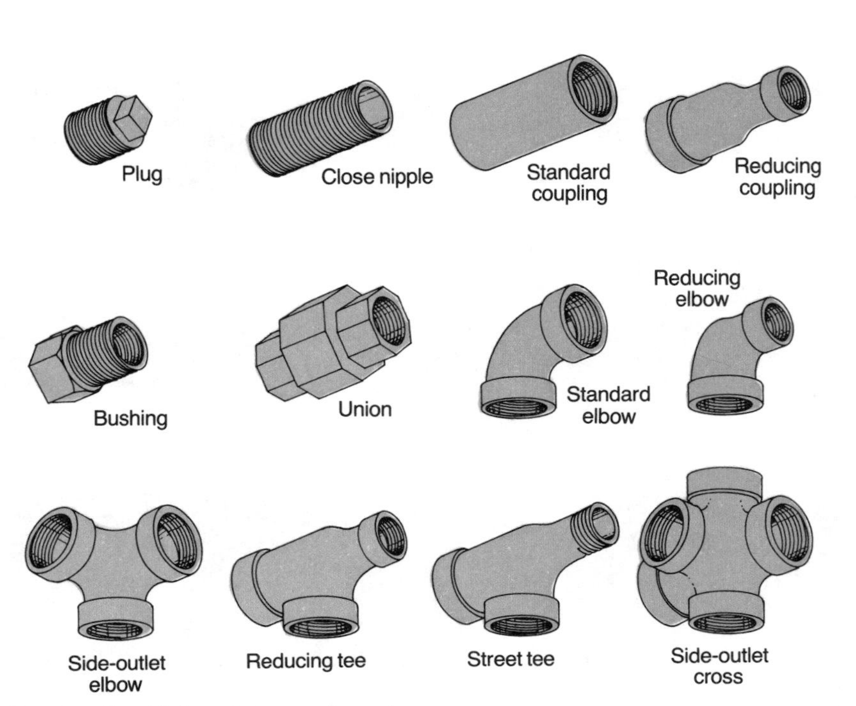

Drawing 18. Fittings for galvanized pipe

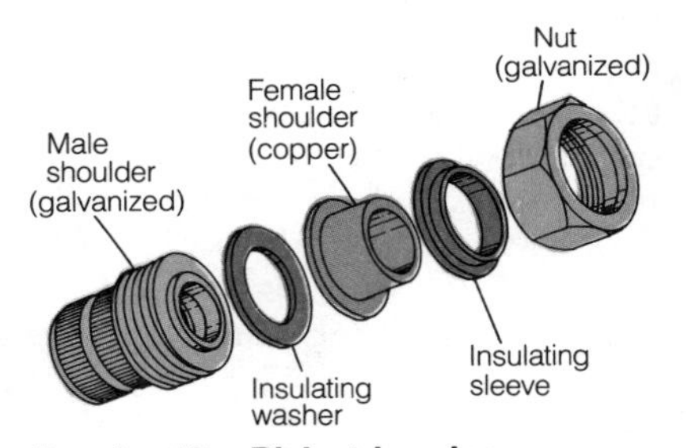

Drawing 19. Dielectric union

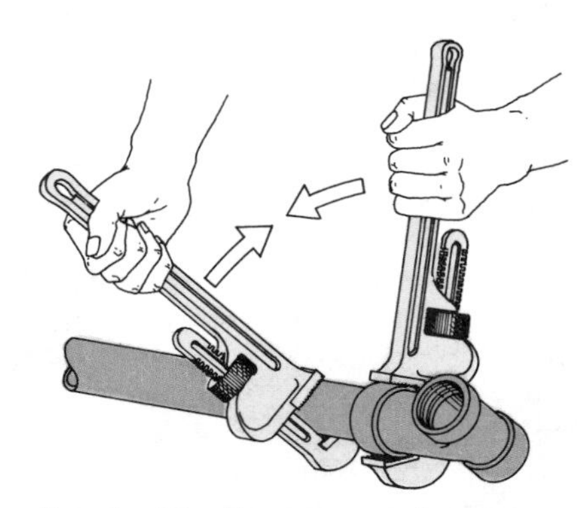

Drawing 20. Use two wrenches at once.

vanized pipe in the same way as cast iron pipe (see page 58).

CAUTION: Before beginning work on supply pipe, turn off the house water supply at the main shutoff valve (see page 23). Drain the pipes by opening a faucet at the low end.

Removing galvanized pipe (supply only). If there's no union in the run, you'll have to saw the pipe in two, because to unscrew the pipe at one end before cutting would be to tighten it at the other end. Hold the run steady with your hand or a wrench to help stop excess motion that would strain joints, and to keep cut ends from sagging. Use a coarse-toothed (18 teeth per inch) hacksaw and place a bucket under the cut to catch any spill.

Applying two wrenches as explained on the facing page in "Using wrenches," unscrew one section of the cut pipe, then the other. If you have trouble, apply liberal doses of penetrating oil to the joints; give it 5 minutes to work into the threads before you begin unscrewing.

Measuring galvanized pipe. To determine how much new galvanized pipe you need, measure the distance between new fittings, then add the distances the pipe will extend into the fittings (see **Drawing 21**). The distance allowed for each fitting is ½ inch for ½ and ¾-inch pipe and ⅝ inch for 1 and 1¼-inch pipe.

Cutting galvanized pipe. It's important to cut galvanized pipe perfectly straight so that threads can be accurately cut in its new ends. Using a pipe cutter with a blade designed for galvanized pipe, follow the directions for cutting copper pipe on page 53. After you've finished cutting, use a knife or the retractable reamer in the cutter handle to remove burrs from the inside of the pipe, and a file to remove burrs from its outside surface.

Threading galvanized pipe

To thread pipe at home you'll need to rent two pieces of equipment: a pipe vise to hold the pipe steady (unless you own a bench vise with special jaws for pipe), and a threader with a head of the same nominal diameter as the pipe. You fit the head of the threader (die) into the threading handle and slip it over the end of the pipe.

To thread the pipe (see **Drawing 22**), exert force toward the body of the pipe while rotating the handle clockwise. When the head of the threader bites into the metal, stop pushing and simply continue the clockwise rotation. Apply generous amounts of cutting oil as you turn the threader. If the threader sticks, some metal chips are probably in the way; back the tool off slightly and blow the chips off.

Continue threading until the pipe extends about one thread beyond the end of the threader's head. Remove the threader from the pipe and clean off the newly cut threads with a stiff wire brush.

Putting galvanized pipe together

The threads on galvanized pipe should be covered with pipe joint compound or fluorocarbon (pipe-wrap) tape to seal them against rust and to make assembly and disassembly easier. Apply pipe joint compound with the brush attached to the lid of the container, using just enough to fill the threads. If you use fluorocarbon tape instead, wrap one and a half turns clockwise around the threads, pulling the tape tight enough so that the threads show through. Do *not* attempt to coat the interior threads of the fitting with compound or tape.

Screw the pipe and fitting together by hand as far as you can. Do this slowly—done too fast, joining creates heat that causes the pipe to expand; later, the pipe may shrink and the joint loosen.

After you've tightened the pipe and fitting by hand, finish with two pipe wrenches as explained previously in "Using wrenches."

Support horizontal runs of new pipe every 6 to 8 feet, vertical runs every 8 to 10 feet; select from the hangers shown in **Drawing 23**.

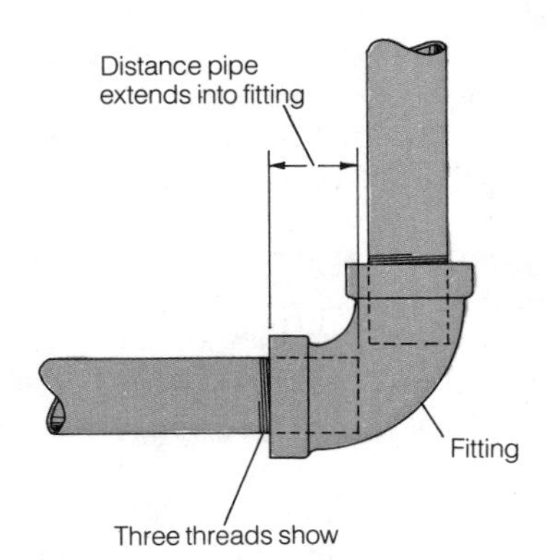

Drawing 21. Screw pipe into fitting; three threads will be visible.

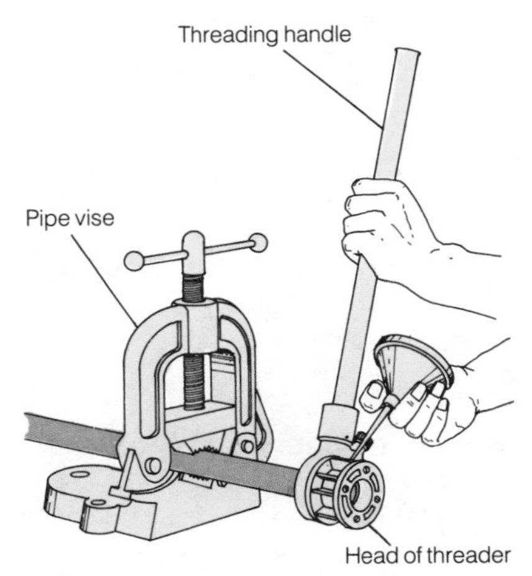

Drawing 22. Thread galvanized pipe with a special vise and a threader.

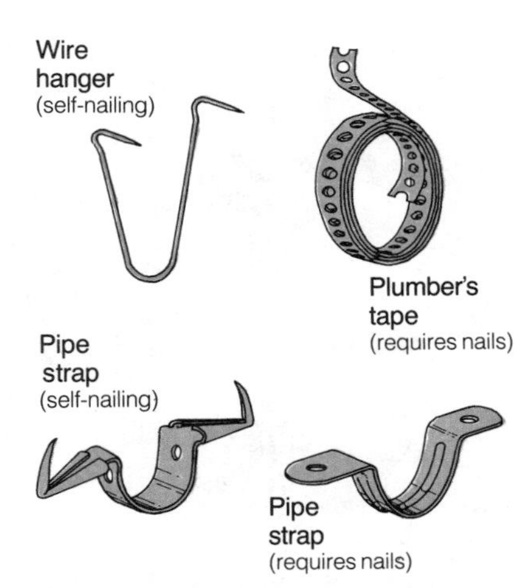

Drawing 23. Supports for runs of galvanized pipe

Working with Cast Iron Pipe

There's a good chance that the drain-waste and vent (DWV) pipe in your house is cast iron. Its fittings put it into one of two categories: the older type is called "hub" or "bell and spigot"; the newer type is called "no-hub" or "hubless." Both types of fittings (see **Drawing 24**) differ from supply fittings in that they have no interior shoulder that could catch waste (see Drawing 8 on page 51).

Cast iron pipe is strong, resists corrosion, and is heavy enough not to chatter when waste flows through it. Since the weight of cast iron makes it difficult to work with, you may want to substitute plastic (see page 50) when repairing or extending cast iron—if your building code allows it. If it doesn't, plan to use no-hub for repairs or extensions: adding it to a run of either hub or no-hub is far simpler than the oakum and lead procedure needed for working with hub fittings.

Like plastic DWV pipe, cast iron DWV pipe is most often sold in lengths of 5, 10, and 20 feet, and in nominal diameters of 1½, 2, 3, and 4 inches, though larger diameters are sometimes available.

Removing cast iron pipe

Because cast iron pipe is so heavy, be sure it's securely supported above and below, or on each side of, the section you need to cut out (see Drawing 32A on page 63). It's sometimes possible to achieve sufficient bracing by wrapping lengths of plumber's tape (see Drawing 23 on page 57) once around the circumference of pipe every 2 feet, then pulling the tape taut and nailing it to nearby joists or studs.

A DWV (drain-waste and vent) cutter (see **Drawing 25**), available at equipment rental stores, makes quick work of removing a length of cast iron pipe. Tightening the knob increases the pressure on the cutting wheels. Pumping the ratchet handle back and forth at right angles to the pipe allows the wheels to dig into the metal, cutting it in two.

Measuring & cutting new cast iron pipe

To determine how much new no-hub pipe you need, simply measure between the cut ends where a section of pipe has been removed.

To cut a length of new pipe, use either a DWV cutter (see **Drawing 25**) or a hacksaw and cold chisel. Using the latter means chalking a cutting line all around the pipe, then scoring it to a depth of 1/16 inch with the hacksaw. Deepen the cut with a soft-headed steel hammer and chisel, tapping around the pipe until it breaks.

Putting cast iron pipe together

To connect a no-hub fitting, or a length of no-hub pipe, to existing cast iron pipe, slip a neoprene gasket (often called a band-aid) over the end of the existing pipe, and the stainless steel band over the end of the new pipe or fitting (see **Drawing 26A**). Bringing the ends together, slip the gasket over the joint, and the band over the gasket (see **Drawing 26B**). Tighten the band with a screwdriver or hexagonal wrench, as required (see **Drawing 26C**).

Support a new horizontal run of pipe at every fitting and every 5 feet, a vertical run every 5 feet. Use plumber's tape as shown in Drawing 23, page 57.

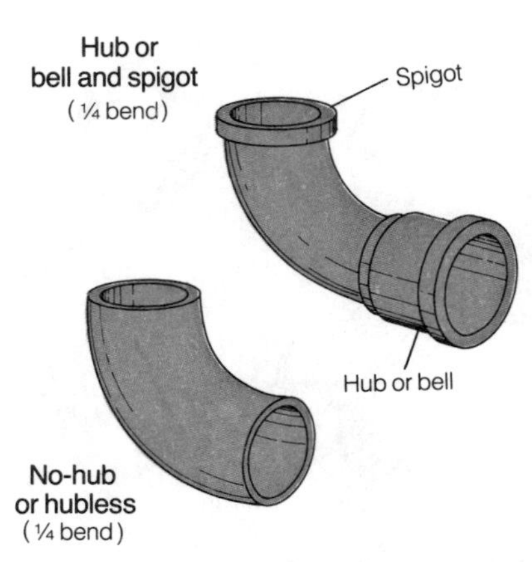

Drawing 24. Two types of cast iron pipe fittings

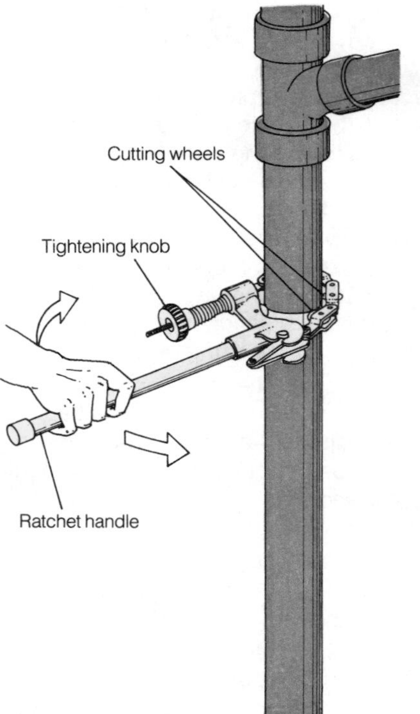

Drawing 25. DWV (drain-waste and vent) pipe cutter

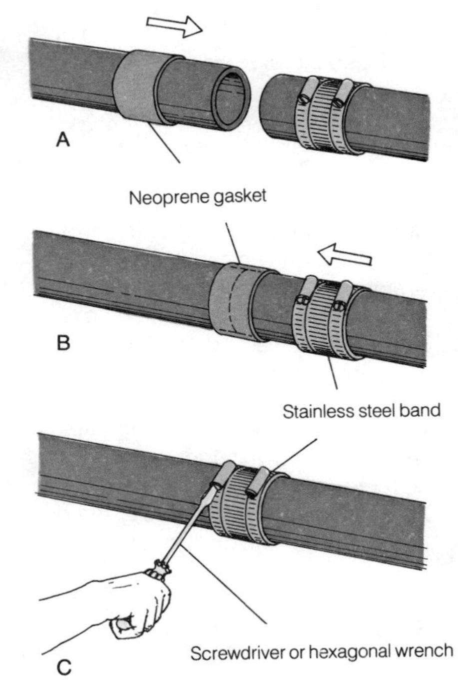

Drawing 26. To join no-hub, slip on gasket, band (A); overlap (B); tighten (C).

Energy conservation tips

Just as dry spells have taught us to save water (for tips see pages 102–109), rising utility costs and shrinking fuel resources have inspired us to conserve energy.

Insulating hot water pipes

To minimize heat loss from your hot water system, insulate all hot water pipes—especially those that pass through unheated or drafty areas. Several types of pipe insulation are available (see **Drawing A**); among the most common are polyethelene foam jackets that fit around most standard pipes and are fastened with tape. Another type of insulation is foil-backed, self-adhesive foam tape, which you spiral-wrap around the pipe.

Making your water heater more efficient

If you live in an average-size house, about 20 percent of your energy bill goes toward heating your water.

Insulating the tank by wrapping it in a special insulating jacket (see **Drawing B**) can save you 5 to 10 percent on your water heating bill. You can buy a kit of fiberglass or foam that you cut to fit around your heater, or buy materials separately and make a jacket from scratch, taping the edges together with electrician's tape.

CAUTION: Keep the insulation away from the pilot light (if there is one) and flue pipe (if there is one). And don't cover the top of a gas heater.

Lowering the temperature setting on your water heater from the average 140° to 110° or 120° will save fuel without making a noticeable difference in laundry or bathing. (Dishwashers, though, require 160° water.)

Remember to turn down the heater during vacations and other periods when your house will be empty.

Installing an automatic timer on your water heater lets you program the thermostat to lower the temperature at low-usage times and raise it at peak-usage times. You can reset or override the timer as hot water needs change.

Adding a heat pump can make a big difference. A heat pump works in conjunction with a conventional electric water heater by using heat from surrounding room air to heat the water. Though energy savings are more substantial in warmer climates and in the summer, payback in energy savings for a heat pump is about 5 years. One drawback: The pump takes twice as long to heat a tankful of water as a conventional electrical heater alone.

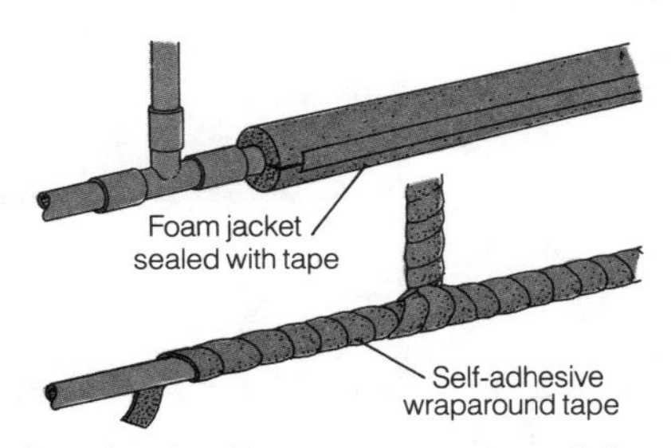

Drawing A. Two common types of pipe insulation

Installing an energy-saving water heater

Most heaters on the market today come with 2 inches of built-in foam insulation—more efficient than fiberglass, which it is fast replacing.

Solar water heaters can harness the sun's energy to supply up to 75 percent of your annual hot water needs (see pages 86–87).

Energy-saving appliances

New models of dishwashers, clothes washers, and other appliances that use hot water have built-in energy saving features such as short wash cycles and low water-temperature settings.

Dishwashers are available with built-in preheating elements that independently heat the water to 160° so that the rest of the plumbing system can operate with cooler water.

Hot water dispensers provide instant steaming water, eliminating the need to boil water on the stove (which takes more energy) for tea, coffee, and soups. Such a dispenser (see page 82) has a small holding tank with a heating element that brings water to 200°.

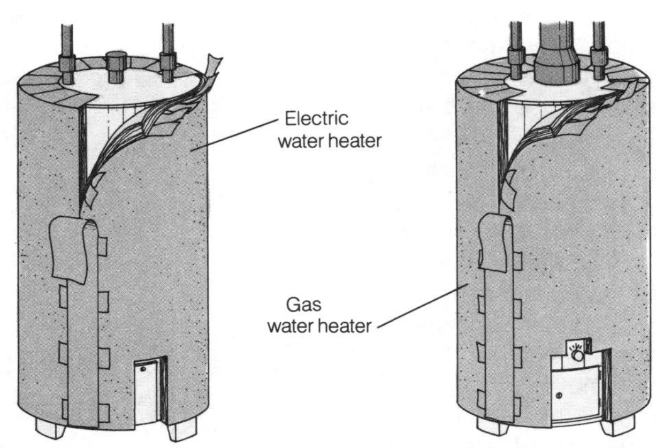

Drawing B. Jackets made of 2-inch-thick R6 fiberglass or ⅜-inch-thick foam insulate either gas or electric water heaters.

Roughing-in & Extending Pipe

How do you go about adding plumbing for a water softener—or an entire new bathroom—to your present system? Start by understanding the basics of plumbing systems (pages 6–11) and pipefitting techniques (pages 47–58). Then review the information in this section—it outlines the planning process, explores some of your options, and presents techniques and general advice for roughing-in the pipes to new fixtures and water-using appliances.

The planning sequence

When plotting out any plumbing addition you must balance code restrictions, the limitations of your system's layout, design considerations, and, of course, your own plumbing abilities.

Check the codes. Don't buy a pipe, a fitting, or a fixture until you've checked your local plumbing and building codes. Almost any improvement that adds pipe to the system will require approval from local building department officials before you start, and inspection of the work before you close the walls and floor.

Learn what work you may do yourself—some codes require that certain work be done only by licensed plumbers.

Map your system. A detailed map of your present system will give you a clear picture of where it's feasible to tie into supply and drain lines, and whether the present drains and vents are adequate for the use you plan.

Starting in the basement, sketch in the main soil stack, branch drains, house drain, and accessible cleanouts; then trace the network of hot and cold supply pipes. Also, check the attic or roof for the course of the main stack and any secondary vent stacks. Determine the materials and, if possible, the diameters of all pipes.

Plan it on paper. Plan the plumbing for any new fixtures in three parts: supply, drainage, and venting. To minimize cost and keep the work simple, arrange a fixture or group of fixtures so they are as close to the present pipes as possible.

Consider these questions, too: Is your present water heater adequate for an additional shower and sink (see page 84)? Is your water pressure adequate for the extra demand (see page 42)?

Decide what kind of pipe you'll need—galvanized, copper, or plastic for the supply pipes; plastic, copper, or cast iron for drain-waste and vent pipes. (For help in deciding, see pages 47–58.)

Layout options. The simplest and most cost-efficient way to add a new fixture or group of fixtures is to connect to the existing main soil stack, either individually or through a branch drain. One common approach is to install a new fixture or group of fixtures above or below a current group on the stack, piggyback style (but check codes carefully). Another plan is to place a new fixture or group of fixtures back-to-back with a present group attached to the main soil stack (see **Drawing 27A**).

If your addition is planned for an area across the house from the existing plumbing, you'll probably need to run a new secondary vent stack up through the roof, and a new branch drain to the soil stack (see **Drawing 27B**) or to the main house drain via an existing cleanout. Installing a new secondary vent stack and branch drain will substantially drive up your labor and demolition costs.

Often a bathroom sink, tub, or shower stall—but not a toilet—can be tied directly into an existing branch drain.

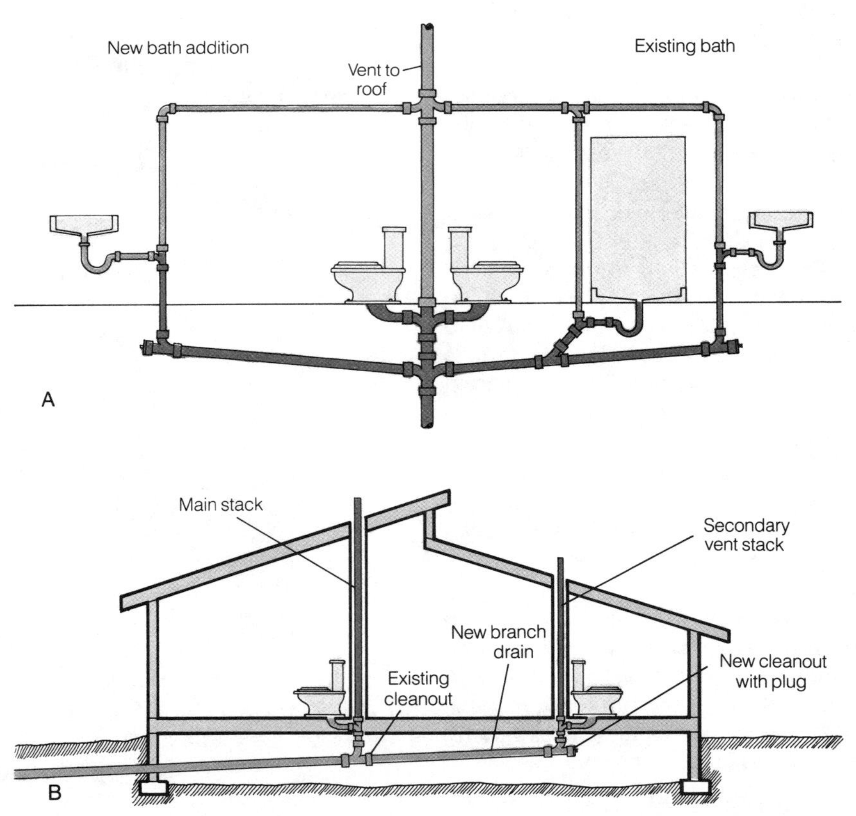

Drawing 27. Plumbing layout options. Two common ways to install a new bathroom are back-to-back with existing fixtures, using the main soil stack (A); and with its own new vent stack and branch drain (B).

Do it yourself? Extending supply, drain-waste, and vent pipes to a new fixture or a new fixture group requires the ability to accurately measure pipe runs, calculate DWV (drain-waste and vent) slope, and cut and join pipe and fittings. In addition, general carpentry skills and tools are needed for opening up walls or floors, notching or drilling framing members, and framing around a bathtub or shower stall. Improvements that involve a new soil stack or the addition of a new venting system are particularly messy and demanding.

If you have any hesitations about these tasks, you could hire a professional to check your plans and install the drain-waste and vent system—or to rough-in all the pipes. Then turn to pages 66–89 and make the fixture or appliance hookups yourself.

A short course on codes

Few code restrictions apply to simple extensions of hot and cold water supply pipes, provided your house's water pressure is up to the task (see page 42). The material and diameter for supply pipes serving each new fixture or appliance should be spelled out clearly in your local code. More troublesome are the pipes that make up the drain-waste and vent (DWV) system—codes govern the organization of these pipes.

Three major elements of the DWV system that come under code restrictions are 1) the vertical *main stack*; 2) horizontal *branch drains*; and 3) separate *vent systems*. When adding a new fixture or a new bathroom, you'll need to answer these questions:

- *Is your present stack* or branch drain adequate in size to tie into?
- *Where can you place fixtures* along your present DWV system?
- *How will each new fixture* be vented? The answers to these questions may make or break your plans.

Sizing stacks, drains, and vents. The plumbing code will specify minimum diameters for stacks and vents in relation to numbers of *fixture units*. (One fixture unit represents 7.5 gallons or 1 cubic foot of water per minute.) In the code you'll find fixture unit ratings for all plumbing fixtures given in chart form.

To determine drainpipe diameter, look up the fixture or fixtures you're considering on the code's fixture unit chart. Add up the total fixture units; then look up the drain diameter specified for that number of units.

Vent pipe sizing criteria also include *length* of vent and *type* of vent, in addition to fixture unit load. (A discussion of vent types follows.)

Critical distance. The maximum distance allowed between a fixture's trap and the stack or main drain that it empties into is called the critical distance. No drain outlet may be completely below the level of the trap's crown weir (see **Drawing 28**) or it would act as a siphon, draining the trap; thus, when the ideal drainpipe slope of ¼ inch per foot is figured in, the length of that drainpipe quickly becomes limited. But if the fixture drain is *vented* properly within the critical distance, the drainpipe may run on indefinitely to the actual stack or main drain.

Venting options. Your four basic venting options (see **Drawing 29**)—subject to local code—are wet venting, back venting, individual venting, and indirect venting.

- *Wet venting* is simplest—the fixture is vented directly through the branch drain or soil stack.
- *Back venting (reventing)* involves running a vent loop up past fixtures to reconnect with the main stack or secondary vent above the fixture level.
- *Individual (secondary) venting,* as the name implies, means running a new vent stack (a "secondary vent stack") up through the roof for a new fixture or group of fixtures distant from the main stack.
- *Indirect venting* allows you to vent some fixtures or appliances (such as a basement shower) into an existing floor drain or laundry tub without further venting.

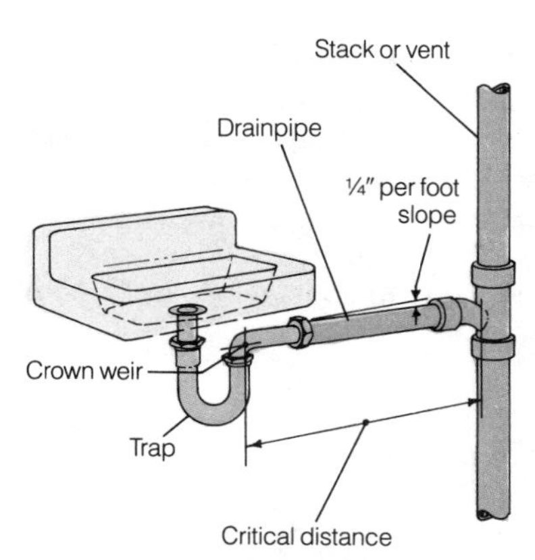

Drawing 28. Critical distance is the maximum drainpipe length allowed between fixture trap and stack or vent.

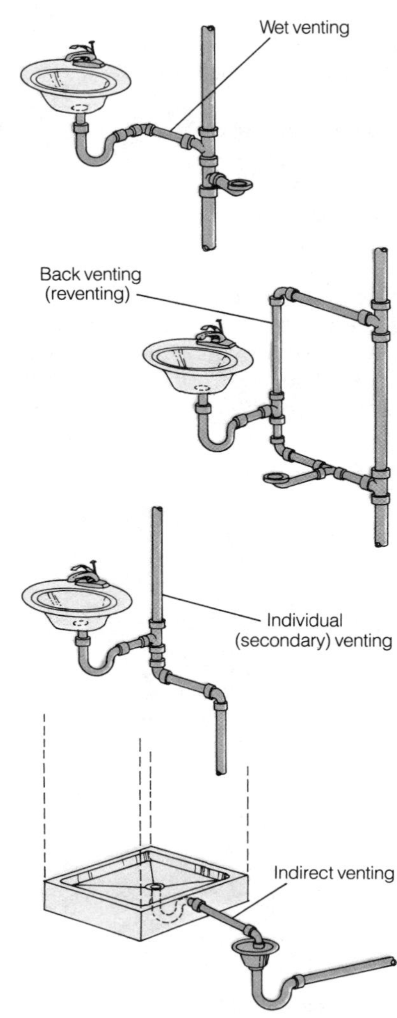

Drawing 29. Four vent types are subject to local codes.

. . . Roughing-in & Extending Pipe

Locating & exposing pipes

Before you can extend the present pipes to reach a new fixture or group of fixtures, you'll need to pinpoint where they run in walls and floors. Then, to gain elbow room for working, you'll need to carefully remove wall, ceiling, and floor materials in the immediate area.

Locating pipe runs. By now, you should roughly know the location of the pipes you'll tie into. Here's where your system map (see page 60) comes in handy.

Locate pipes as exactly as possible from above or below. You may have to drill or cut exploratory holes to pinpoint the location of a stack, branch drain, or supply risers inside a wall or ceiling. Once you find one riser, the other should be about 6 inches away.

Exposing pipes in a wall. To open the wall, insert a metal tape measure into the exploratory hole near the pipes you're tying into. Run the tape left until you hit a stud; note the measurement, then mark the distance on the outside of the wall covering. Then repeat the process to the right. With a carpenter's level that shows plumb also, draw vertical lines through your marks to outline the edges of the flanking studs. Then turn the level to the horizontal and connect the vertical lines above and below where you plan to tie into the pipes. A rectangle about 3 feet high should be large enough.

To cut into gypsum wallboard, drill small pilot holes at the four corners of your outline, then use a keyhole saw to cut along the lines you marked (see **Drawing 30**).

Exposing pipes in floors and ceilings. Floors can be a far messier proposition than walls—you have to tear out not only the floor covering (and repair it later), but also the subfloor beneath. If you're tying into a branch drain, try to gain access from below—from the basement for a first-floor drain, or through ceiling materials below the second floor. To open a ceiling between joists, follow the procedure for opening walls (preceding).

Pipe connections

Basically, tying into drain-waste, vent, and supply lines entails cutting a section out of each pipe, inserting a new sanitary fitting or supply tee, and running pipes to the new fixture along preplotted lines.

Laying out the plan. Most new fixtures will include a "fixture template" or other roughing-in measurements telling where supply pipes and the trap exit into the drainpipe (the spot where the drainpipe enters the wall or floor) should be located on the wall or floor. Position these measurements carefully on the wall or floor where you prefer them. The combined *length* of the new fixture's drain and the *height* of its trap exit on the wall or below the floor will determine exactly where the connection to stack or branch drain will be made.

To plot a stack connection inside a wall (for a sink), first mark the fixture's roughing-in measurements on the wall material. To plot the ¼-inch-per-foot slope for your drainpipe, run a tape measure from the center of the trap exit mark to a point at the same height on the stack. Subtract ¼ inch per foot of this distance, and lower the mark on the stack by this amount (see **Drawing 31**).

Position the new closet bend (for a toilet) or the trap (for a shower stall or bathtub) below the subflooring. Figure slope with a chalk line snapped on a parallel joist, or a string pulled taut along the proposed run.

Supply pipes are not required to slope ¼ inch per foot as drainpipes are, but figuring in at least a slight slope allows you to drain the pipes later. Run hot and cold supply pipes parallel to each other, at least 6 inches apart.

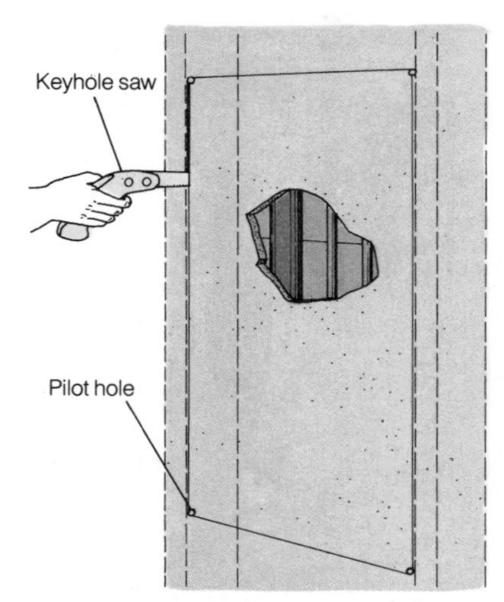

Drawing 30. Cut into gypsum wallboard with a keyhole saw to get to stack and other pipes inside wall.

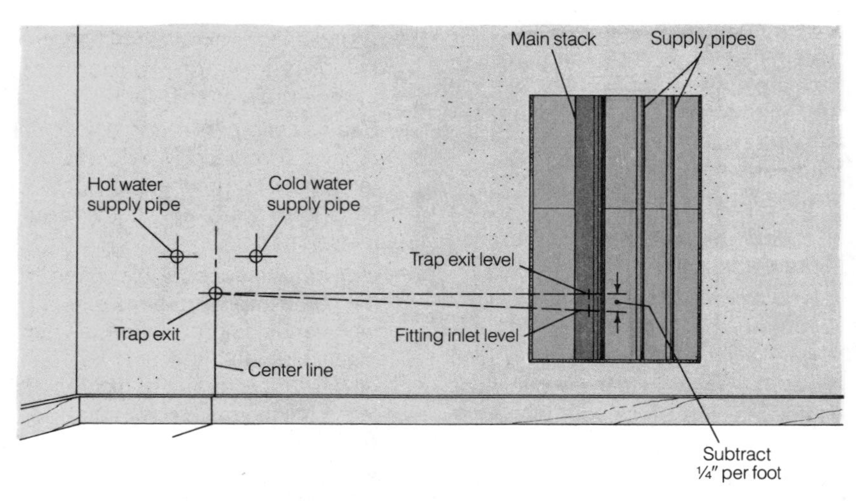

Drawing 31. Plot slope for a sink drainpipe by measuring from center of trap exit mark to a point at same height on stack. Subtract ¼ inch per foot of distance, then mark stack at lower point. New mark gives correct slope.

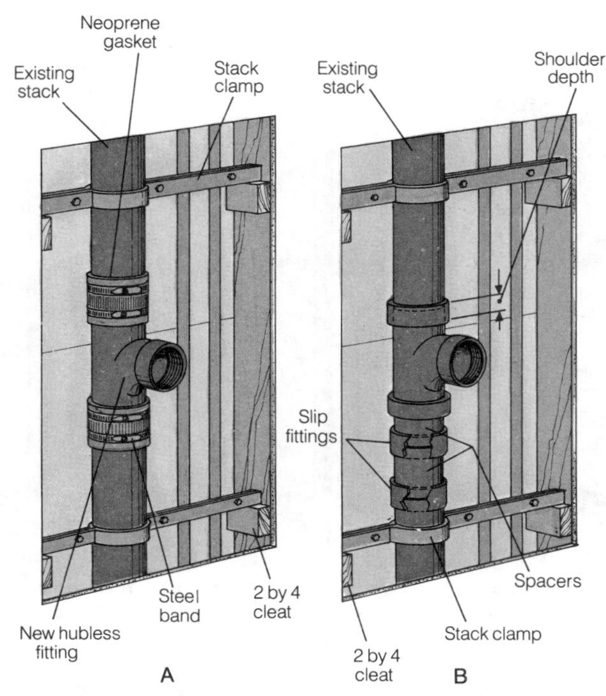

Drawing 32. To connect new drain-waste and vent (DWV) fittings, use neoprene gaskets and steel bands with a cast iron stack (A); or solder or cement fittings to copper or plastic (B).

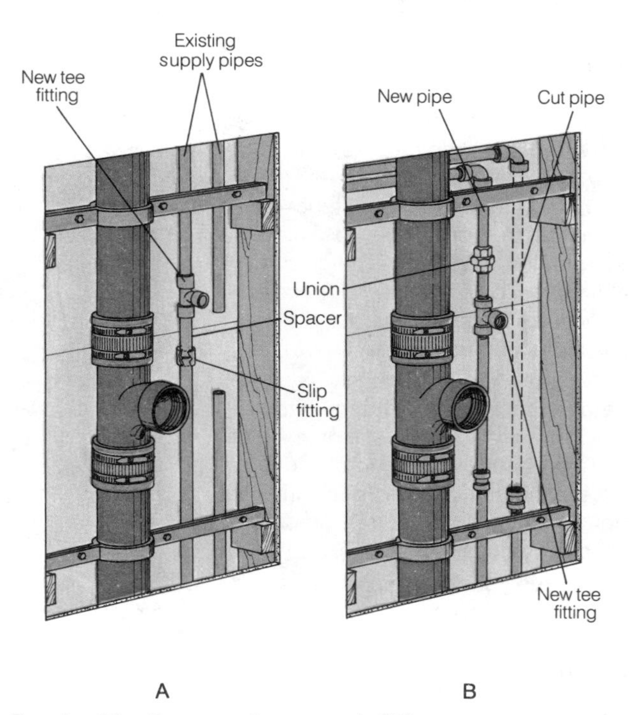

Drawing 33. To connect new supply fittings, use spacers and slip fittings with rigid plastic or copper pipe (A); follow galvanized pipe back to nearest fitting on each end (B).

Drain-waste and vent connections. The method of tying into DWV pipes depends on pipe material.

- *For the common cast iron stack.* Hold your new hubless fitting (see page 58) with its inlet centered at the mark where the drainpipe will enter the stack (see "Laying out the plan," preceding); mark the top and bottom edges of the new fitting on the stack. Before cutting into a stack, support it by installing stack clamps (see **Drawing 32A**) several inches above and below your cutting marks. Nail 2 by 4 cleats to the sides of the studs; then slip the clamps into position and tighten the bolts.

  Cut the stack at the marks with a DWV cutter (see page 58). Slip a neoprene gasket over each end of the cut pipe and position the new fitting. Then slide the gasket over each joint and screw the steel bands semitight. (You can go back later to readjust and tighten.)

- *For plastic or copper stacks.* First mark the fitting on the stack, as described previously. Then mark the depth of the new fitting's *shoulder* below the top mark, and make another mark 8 inches below the fitting's bottom mark. Carefully, with a hacksaw, cut the stack at this second set of marks.

  Solder or cement the new fitting to the top pipe, cut a short spacer as shown, and join it to the bottom of the fitting. Measure another spacer that fits the remaining space exactly. Slide two slip fittings onto the pipe, position the spacer, and solder or cement the two fittings over the joints (see **Drawing 32B**). Or, in place of slip fittings, use neoprene gaskets and steel bands with a plastic stack.

- *The connections to a branch drain* are accomplished with the same techniques outlined above (depending on material), but the pipes are horizontal. Be sure the branch drain is supported adequately with pipe hangers on each side of the cut (see Drawing 23 on page 57).

Supply connections. Connections to supply risers are made in the same way as connections to drainpipes. First, shut off the water supply at the main shutoff valve (see page 23) and drain the pipes, if possible.

The exact method of connection varies with material. If your supply pipes are soft copper or flexible plastic, just cut the lines and insert new tee fittings. Rigid plastic or hard copper require that you cut out a section of pipe—about 8 inches—and, depending on the play available, add one or two spacers (nipples) and slip fittings (see **Drawing 33A**). Or, for plastic pipe, install a spacer with one threaded end and a union.

If your supply pipes are threaded galvanized pipes, you'll have to cut each pipe (pages 56–57), then follow it back to the nearest fitting on each end (see page 56).

Unscrew the pipe from each fitting, using two wrenches (see Drawing 20 on page 56). Either install a union, the new pipe, and the tee fitting (see **Drawing 33B**), or—if you wish to change to plastic or copper at this point—add a transition fitting, a spacer, the new pipe, and a tee fitting.

. . . Roughing-in & Extending Pipe

New pipe runs

With the connections made, the new DWV and supply pipes are run to the new fixture location, as marked earlier. If you wish to change pipe type—say from cast iron or copper to plastic—it's a matter of inserting the appropriate adapter at the fitting end (transition fittings are described under the various types of pipe, pages 47–58).

Ideally, pipes should always run parallel to framing members, and between them. Actually, at some point you'll probably have to employ one of the techniques detailed below.(Note: Before cutting any joists or studs, check your local building code.)

Leave your new pipe fittings and runs uncovered for a few days to check for leaks. Then patch the wall, ceiling, or floor.

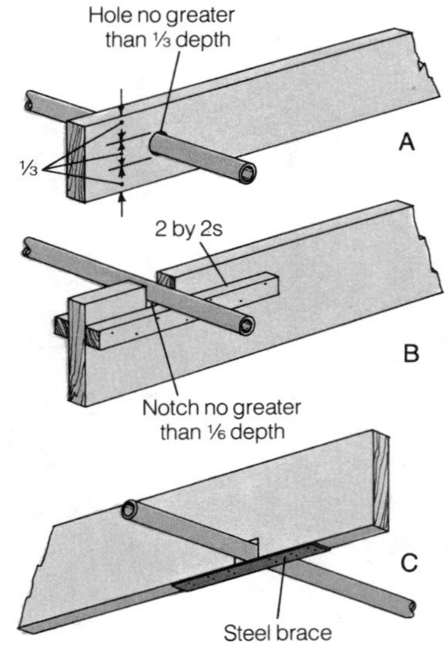

Drawing 34. To run pipe through a joist, drill a hole (A); notch the top (B); or notch the bottom (C).

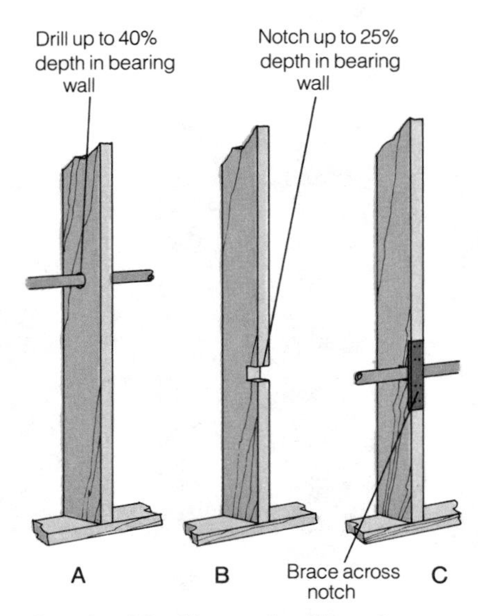

Drawing 36. To run pipe through a stud, drill a hole (A) or notch the edge (B) and brace notch with steel ties (C).

Cutting into joists. If a pipe hits a joist near its center, you can normally drill a hole (see **Drawing 34A**), as long as the diameter is no greater than one-third the depth of the joist.

If a pipe run hits a joist near its top or bottom, a notch may accommodate it. The depth of the notch must be no greater than one-sixth the depth of the joist, and the notch cannot be located in the middle third of the span. Top-notched joists (see **Drawing 34B**) should have lengths of 2 by 2 wooden cleats nailed in place under the notch on both sides of the joist to give added support. Joists notched at the bottom (see **Drawing 34C**) should have either a steel tie or a 2 by 2 cleat nailed on.

You might sometimes need to cut an entire section out of a joist to accommodate a drain-waste and vent (DWV) section. Make this cut only in the end quarters of the joist. Reinforce that section by using headers on both sides of the cut (see **Drawing 35**).

Cutting into wall studs. You can drill a hole that's up to 40 percent of the stud depth (see **Drawing 36A**) in bearing walls (those that support joists or rafters), and up to 60 percent in nonbearing walls—or bearing walls if you nail another stud to the first stud for strength. Holes should be centered.

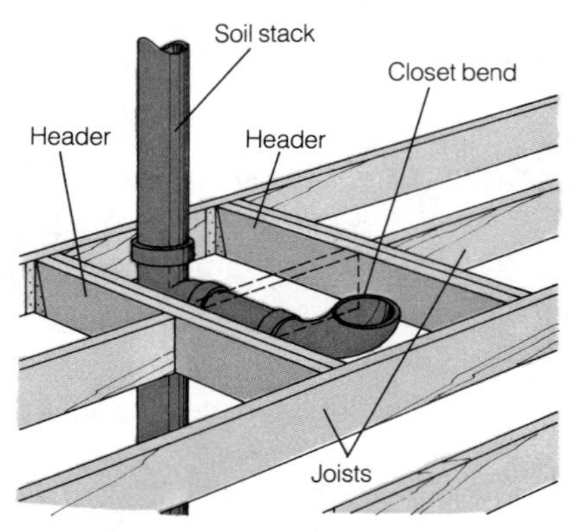

Drawing 35. If you remove a joist section to run DWV pipe, reinforce cut with headers on both sides.

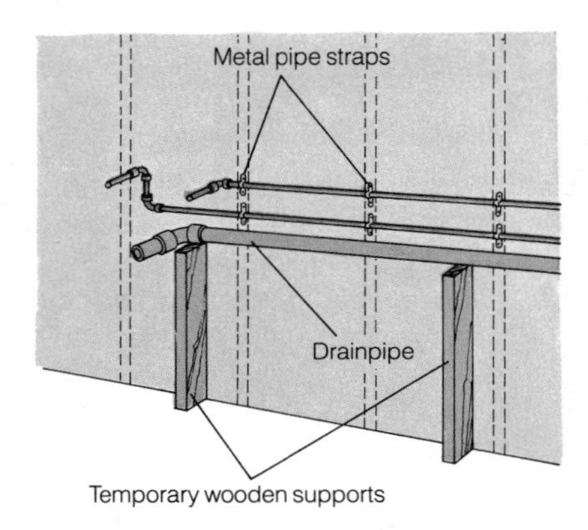

Drawing 37. To run exposed pipes along a wall, attach pipe hangers (such as pipe straps), then remove supports.

You can notch up to 25 percent of the stud depth in bearing walls (see **Drawing 36B**) and up to 40 percent in nonbearing walls. The notches should be braced with steel ties (see **Drawing 36C**).

Running pipes outside the wall. This requires no notching and leaves less wall that needs to be patched. Temporarily support pipes with blocks of wood; then fasten pipe hangers to wall studs (see **Drawing 37**). You can hide the pipes by building cabinets, a closet, a vanity, or shelves over them.

"Thickening" a wall. New studs and wall materials may be erected in front of the old wall, both to cover pipes hung on the wall and to accommodate oversize drain-waste and vent (DWV) fittings, such as a new stack. Build out the entire wall, or thicken the lower portion only, leaving a storage ledge or shelf above the pipes.

Building up the floor will cover a new branch drain. Either thicken the floor with furring strips around the new pipes, or build a platform over the pipes for the fixture or appliance.

Roughing-in fixtures

Following are general tips, installation notes, and precautions for roughing-in new fixtures that require tying into your present drain-waste and vent (DWV) and supply systems, or extending them.

A bathroom sink is comparatively easy to install (see **Drawing 38A**). Common installations are back-to-back (requires little pipe), within a stock vanity cabinet (hides pipe runs), and side-by-side (see page 71). A sink can often be wet-vented if it's within the critical distance; otherwise it's back-vented. Adding a sink has little impact on the drain's present efficiency (a sink rates low in fixture units).

Pipes required: Hot and cold supply stubouts; shutoff valves; transition fittings, if necessary; flexible tubing above shutoff valves. Air chambers may be required.

Toilet. The single most troublesome fixture to install, a toilet (see **Drawing 38B**) requires its own vent (2-inch minimum) and at least a 3-inch drain. If it's on a branch drain, a toilet can't be upstream from a sink or shower.

The closet bend and toilet floor flange must be roughed-in first; the floor flange must be positioned at the level of the eventual finished floor.

Pipes required: Cold water supply stubout with shutoff valve; flexible tubing above valve. One air chamber may be required.

Shower stall and bathtub. Like a sink, bathtubs and showers rate low in fixture units; they're often positioned on branch drains and are usually wet-vented or back-vented; both enter the stack at floor level or below because of the floor drain trap. A shower's faucet body and shower head assembly are installed while the wall is open (see **Drawing 38C**); tubs and showers may require support framing.

Pipes required: Hot and cold supply lines and a pipe to a shower head. Use flexible tubing above supply stubouts. Air chambers may be required.

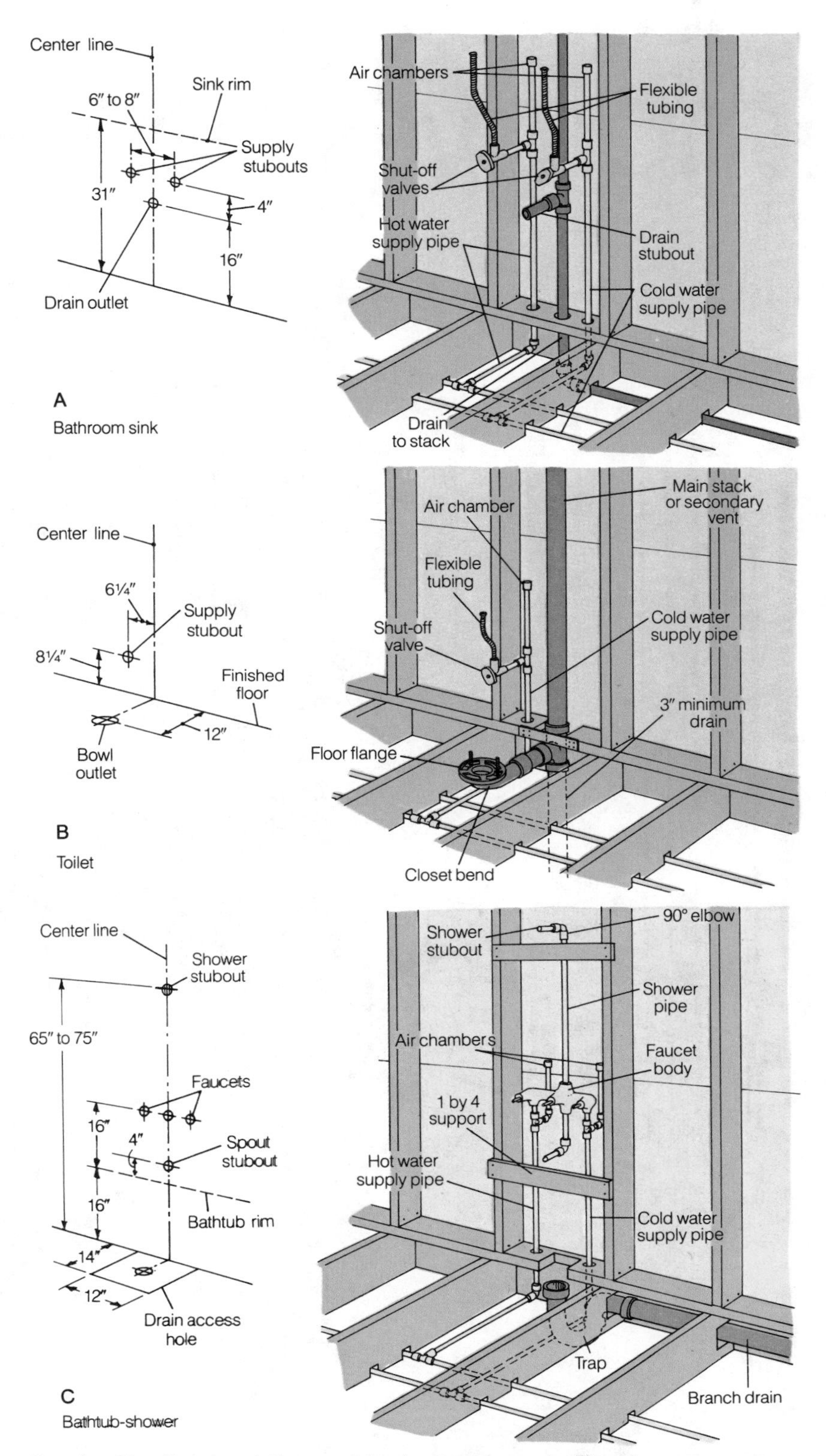

Drawing 38. Representative roughing-in measurements. Plumbing components illustrated are components of a new bathroom: a sink (A), toilet (B), and combination bathtub-shower (C). Use the measurements to help you plan; check local codes and specific fixture dimensions for exact roughing-in requirements.

Plumbing Improvements

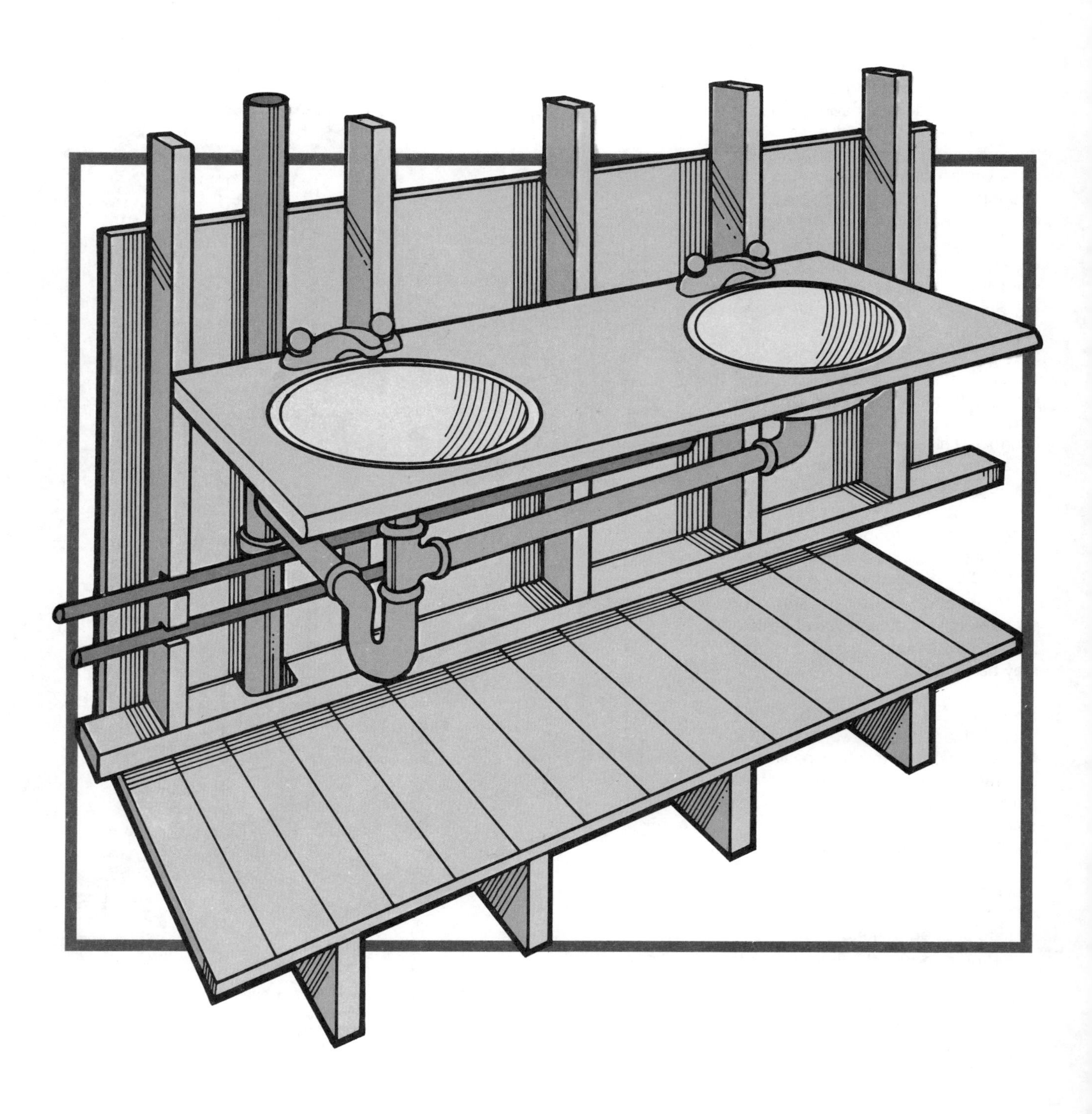

Replacing Wall-mount Faucets

Time to replace a beyond-repair faucet? The job is only a little more difficult than some repairs—and well worth the effort. Often removing the old fixture takes longer than installing the new.

Most kitchen and bathroom faucets today are deck mounted, meaning that they are attached to the sink or countertop through precut holes. But some faucets, especially those in older homes, are mounted directly on supply pipe stubouts above the sink instead of on the sink itself. These are called wall-mount faucets. They are either compression faucets (pages 13–14) or washerless faucets (pages 15–18).

Whatever the type of faucet, converting from a wall-mount to a deck-mount faucet is a big job that involves rerouting pipes, patching walls, and more. You'd be better off sticking with a wall-mount—choosing an updated style—rather than switching over to a deck-mount faucet.

Selecting a wall-mount faucet

Before shopping for a new faucet, measure the center-to-center distance between the water supply pipes—usually 4, 6, or 8 inches. Also, carefully measure the diameter of the supply pipes. Choose a faucet as close as possible to the old one in center-to-center distance—take the old faucet to the store with you if you can.

You'll find a number of brands to choose from. Look for a replacement unit of good quality, made by a reputable company. It should come with clear, complete installation instructions. You should also find out whether the manufacturer offers repair kits and replacement parts for the faucet.

Removing a wall-mount faucet

When removing either type of wall-mount faucet (compression or washerless); use a tape-wrapped wrench to free up connecting fittings. This prevents the wrench teeth from marring or stripping the fittings.

CAUTION: Before doing any work, turn off the water at the fixture shutoff valves or the main shutoff valve (see page 23). Open the faucet to drain the pipes.

Removing a wall-mount compression faucet. Take the handles off as described on page 13. Using the tape-wrapped wrench, unscrew the chrome-plated coupling nuts (see **Drawing 1**) on the faucet body behind the faucet handles. Pull the faucet body off the shanks projecting from the wall.

Removing a wall-mount washerless faucet. Using the tape-wrapped wrench, turn the spout ring counterclockwise until the spout lifts off. Remove the escutcheon to get at the connections. If the escutcheon won't budge, apply some penetrating oil and try again. Set aside the escutcheon and unscrew the connecting nuts (see **Drawing 2**) that join the faucet body to the water supply pipes.

Hooking up a new wall-mount faucet

Just about any plumbing fixture you buy will come with specific installation instructions. Units differ from one manufacturer to another, so you'll need to read and follow instructions carefully.

Be sure to put all gaskets and washers in place before making connections. To ensure a leakproof hookup, apply pipe joint compound (see **Drawing 3**) or wind pipe-wrap tape around all pipe threads.

Test the faucet for leaks, tightening any loose connections.

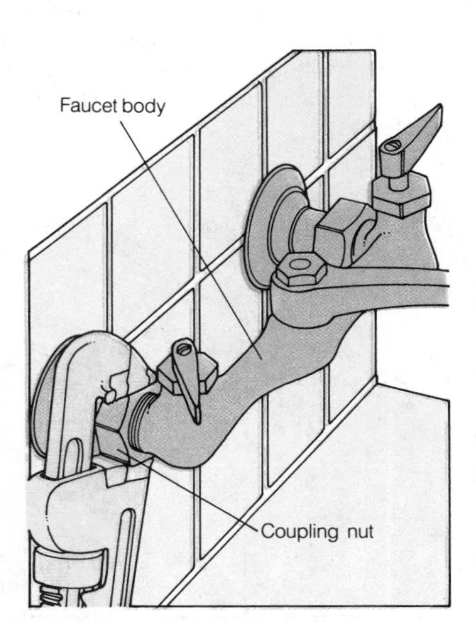

Drawing 1. To remove a wall-mount compression faucet, unscrew the coupling nuts.

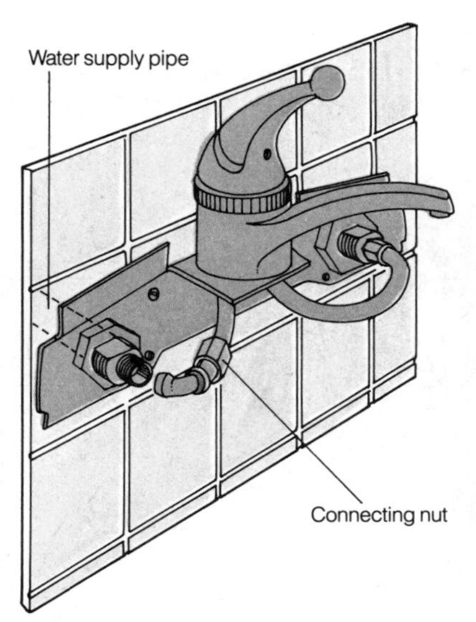

Drawing 2. To remove a wall-mount washerless faucet, loosen the connecting nuts on the faucet body.

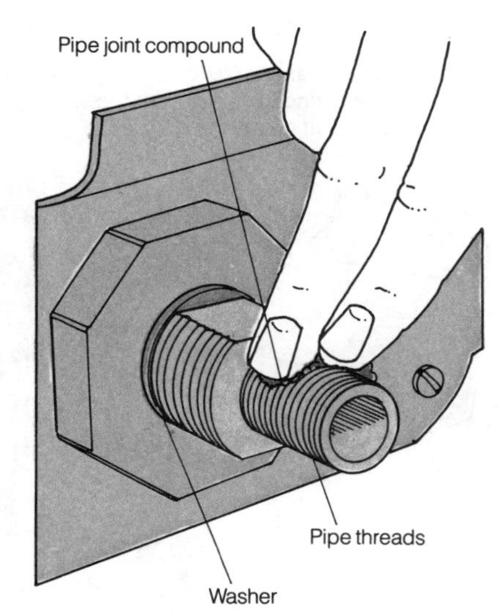

Drawing 3. Before hooking up a new faucet, put all washers and gaskets in place; apply pipe joint compound to threads.

Replacing Deck-mount Faucets

When shopping for a new deck-mount faucet, you'll find the selection staggering. You can choose from an assortment of single-handle washerless faucets—valve, disc, ball, cartridge—and from antique-style reproductions to futuristic compression faucets. All are interchangeable as long as the faucet's inlet shanks are spaced to fit the holes of the sink you'll be mounting it on.

If possible, take the old faucet along with you when you buy a replacement. Also measure the diameter of the supply pipes.

Choose a new unit that comes with clear installation instructions and has repair kits or replacement parts available for future use.

CAUTION: Before doing any work, turn off the water at the fixture shutoff valves or the main shutoff valve (see page 23). Open the faucet to drain the pipes.

Working with flexible tubing. Some faucets come with the copper or plastic flexible tubing for the water supply already attached. Plain copper flexible tubing is most often used where it will be concealed by a cabinet. If the tubing will be in plain view, or the existing tubes are damaged or scratched, you might want to buy new chrome-plated copper tubes, or plastic ones, along with the faucet.

Since the lengths of tubing are flexible, you can bend them to get from the faucet inlet shanks to the shutoff valves on the stubouts at the wall. If necessary, use a hacksaw or small pipe cutter to cut the tubing to the correct length.

If the faucet doesn't come with flexible tubing attached, fasten the new or existing flexible tubing first to the shutoff valves under the sink and then to the faucet inlet shanks.

For more information about flexible tubing, see pages 52 and 71.

Removing a deck-mount faucet. Since space is limited under the sink, use a basin wrench to remove the couplings connecting the flexible tubing to the faucet inlet shanks (see **Drawing 4**). Still using the basin wrench, loosen the locknuts on both shanks and remove the locknuts and the washers. Then lift out the faucet.

Disassembling a pop-up. If you're working on a bathroom sink that has a pop-up assembly, remove the stopper (see page 21). If it's connected to a pivot rod, disengage the pivot rod (see **Drawing 5**) by unfastening the clevis screw and spring clip. Lift out the pop-up rod.

Disconnecting a sink spray. On a kitchen sink with a spray hose attachment, use an adjustable or basin wrench to undo the coupling connecting the hose to the hose nipple under the faucet body (see **Drawing 6**).

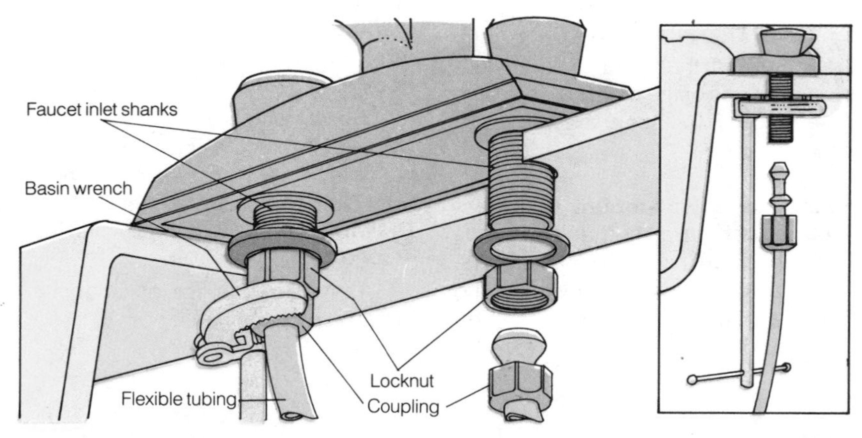

Drawing 4. Unfasten the coupling from each of the faucet inlet shanks, using a basin wrench and working from beneath the sink. Then remove the locknuts and washers from the shanks and lift out the faucet.

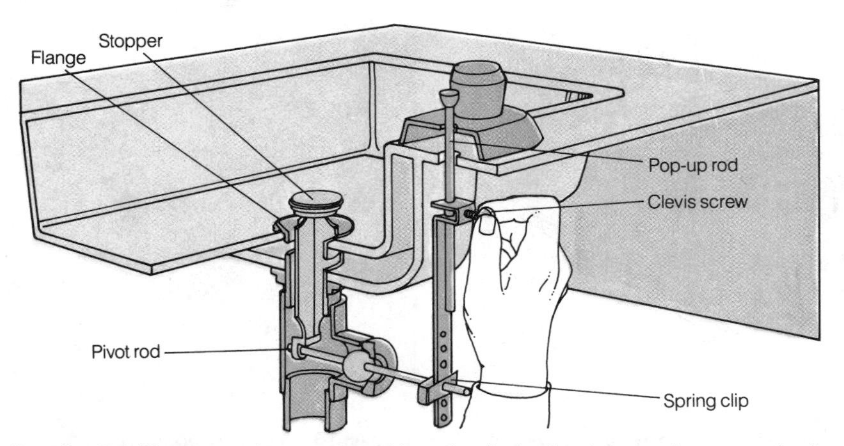

Drawing 5. Disengage the pop-up rod from the pivot rod and stopper by unscrewing the clevis screw and releasing the spring clip.

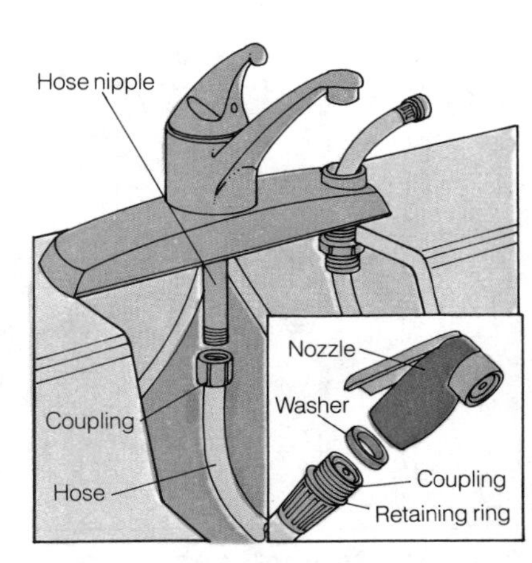

Drawing 6. Remove the spray hose from the hose nipple under the faucet body by detaching the coupling.

Installing a new faucet. Wipe off the surface of the sink where the new faucet will sit. Most faucets come with a rubber gasket on the bottom; if yours doesn't have one, apply plumber's putty (see **Drawing 7**).

Set the faucet in position, simultaneously feeding the flexible tubing (if attached) down through the middle sink hole (see **Drawing 8**). Press the faucet onto the sink's surface. Screw the washers and locknuts onto the faucet inlet shanks by hand; tighten further with a tape-wrapped basin wrench. If your sink has a spray hose, attach it next (see page 19).

Connect the flexible tubing (see **Drawing 9**), gently bending the tubes to meet the shutoff valves. (If there are no shutoffs, or if pipe extensions are needed, install them as described on pages 70–71.) Join the tubing to the shutoff valves, using compression or flared fittings (pages 48 and 55).

Drawing 7. Apply plumber's putty to the bottom edge of the faucet if there is no rubber gasket to seal it to the sink's surface.

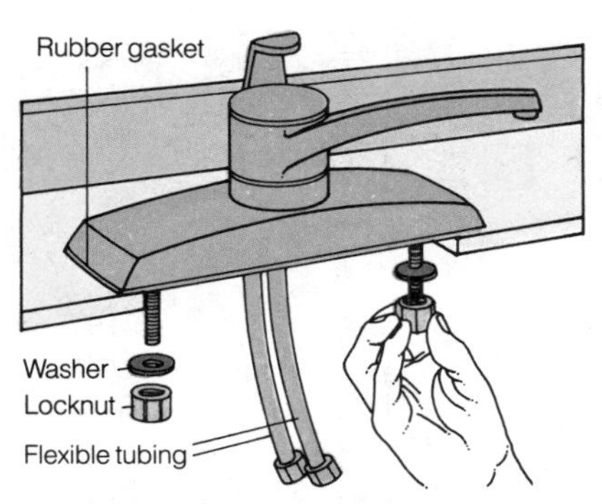

Drawing 8. Set the faucet in place, threading the supply tubes through the sink hole and press assembly onto the sink.

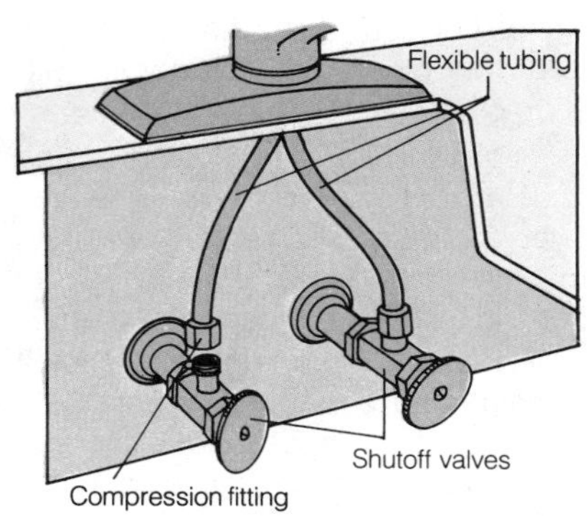

Drawing 9. Attach the flexible tubing to the shutoff valves, using compression or flared fittings.

Caring for new fixtures

Whether made of stainless steel, porcelain enamel, or fiberglass, new fixtures can continue to look as good as new.

Stainless steel fixtures stay bright and shiny if given light and frequent cleaning with a damp cloth, followed by drying with a soft cloth.

For a high polish, apply a mildly abrasive cleanser. For difficult spots and stains that aren't removed by daily cleaning, try ammonia in water, or use a solvent like alcohol, baking soda, vinegar, or turpentine applied with a rag. Follow any of these applications with detergent and hot water; then rinse and dry with a soft, clean cloth.

Drawing A. To remove stains on a porcelain enamel sink, pour bleach into hot water and let soak until stain comes off.

Porcelain enamel fixtures should be cleaned gently with sudsy water, followed by a rinse and dry. For stubborn stains, fill the sink or tub with hot water and add chlorine bleach or oxygen-based bleach (see **Drawing A**) diluted according to label instructions. Let this stand and soak until the stain can be rubbed off. Full-strength or improperly diluted bleach can damage the surface. Abrasive cleansers can also do harm—they contain gritty substances that eventually wear away the enamel surface, making it more difficult to clean as time goes on.

For toilets, use a chlorine bleach or a special toilet cleaner. Foaming spray cleaners for the bathroom and liquid cleaners also offer an easy and effective way to clean without damaging a fixture's surface.

Fiberglass fixtures are coated with a protective gel sealant that wears off in time. You can postpone this dulling by cleaning with liquid cleaners only. Abrasive cleansers should never be used on fiberglass—they will damage the surface. If this happens, the original finish can be restored by the manufacturer of the unit.

Apply liquid automobile wax or a special fiberglass polish from time to time to maintain a protective shine.

Adding Shutoff Valves

If your house is not outfitted with shutoff valves, or if you have a valve that's worn, you'll find that installing a new one is a not-too-difficult task and one that will make future repairs less of a hassle.

A shutoff valve simplifies turning off the water supply to a fixture for repairs or in an emergency—you just turn the valve handle clockwise until it's fully closed, open the faucet(s) to drain the pipes, then go ahead with your job.

Take a quick survey of your home to determine your shutoff valve needs. Every sink, tub, shower, and washing machine should have shutoffs for both hot and cold water pipes. A toilet and a water heater require only one shutoff valve (because they use only cold water), and a dishwasher needs only one shutoff valve because it uses only hot water.

CAUTION: Before doing any work, turn off the water at the main shutoff valve (see page 23). Open the faucet to drain the pipes.

Selecting a shutoff valve. When you shop for a shutoff valve, you'll need to choose either an angled valve or a straight one. Angled valves are used when the supply pipe, called a stubout, comes from the wall; straight valves are used for pipes that come up from the floor. Choose a globe valve instead of a gate valve (see page 22)—it's more reliable, it's easier to repair, and, unlike a gate valve, it can control the amount of water flow.

There's a selection of materials, too. Select a shutoff valve that fits the existing pipe and is compatible in material. Copper tubing takes brass valves; iron and plastic pipe use iron and plastic valves, respectively. You can use transition fittings if the pipes aren't compatible; these fittings allow you to change materials of the valve and its connecting pipe (from galvanized iron to plastic, for example).

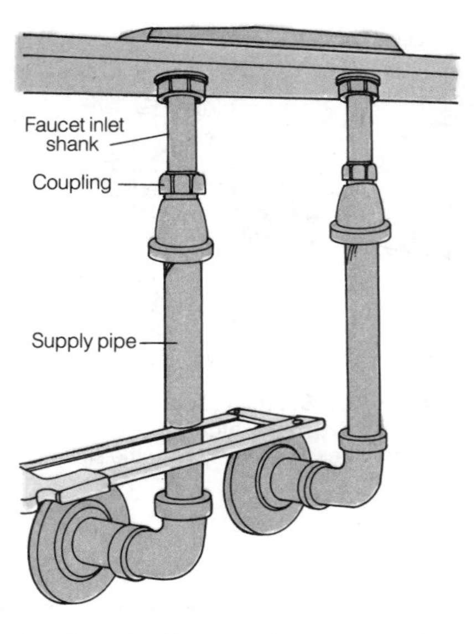

Drawing 11. Use a hacksaw to cut out a section of the existing supply pipe, then free the supply pipe from the faucet inlet shank.

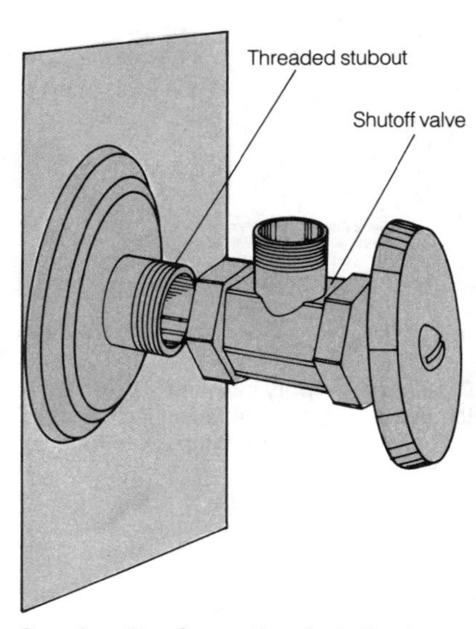

Drawing 13. Screw the shutoff valve directly to a threaded stubout or onto a transition fitting.

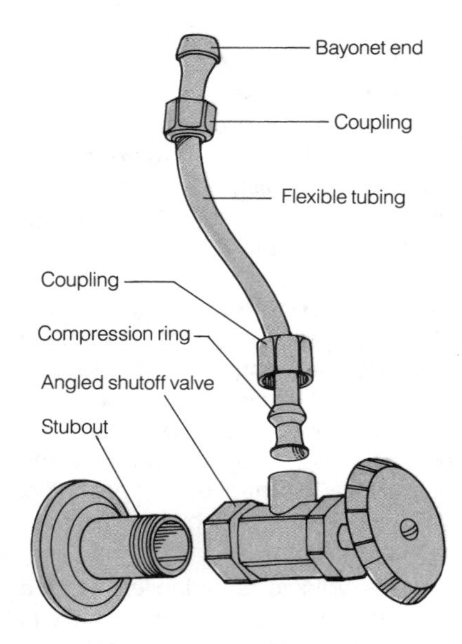

Drawing 10. Use flexible tubing to hook up the shutoff valve for a sink, toilet, tub, or shower.

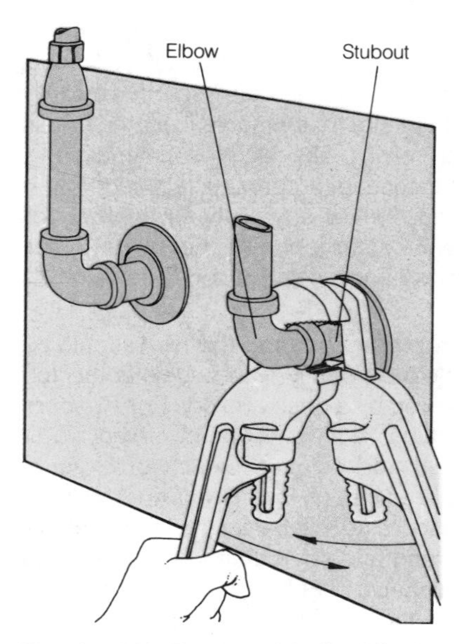

Drawing 12. Remove the elbow from a galvanized stubout, using the opposing force of two tape-wrapped wrenches.

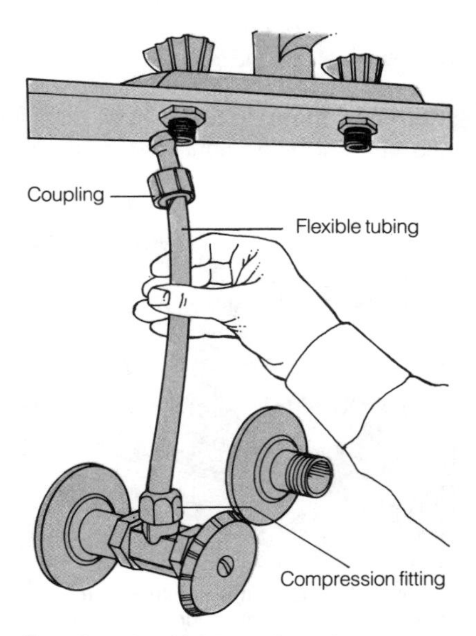

Drawing 14. Hook up the valve to the fixture by tightening the coupling and the fitting on the flexible tubing.

You may want a chrome-finished valve if it will show.

Besides all of that, the kind of stubout you have dictates the kind of fittings you need. A threaded pipe naturally requires a threaded valve; a copper stubout takes a fitting that's soldered (see pages 53–55) at one end and threaded at the other.

Selecting flexible tubing. Lengths of flexible tubing (see **Drawing 10**), sometimes called flexible connectors, save you the trouble of piecing pipe together to join the valve to the fixture. These plain copper, chrome-plated copper, or plastic tubes are available in a variety of lengths, so calculate the exact size you need.

Flexible tubing for a sink or basin has a cylindrical rubber gasket on one end and is ½ inch in diameter; tubing for a toilet has a large conical gasket on one end and a diameter of 7/16 inch.

Unless you buy a kit that contains a shutoff valve and tubing already joined, you must join them yourself with a flared connection (see pages 48 and 55) or a compression fitting (see pages 48 and 55). If you opt for the kit, you'll get a valve, flexible tubing, fittings, and instructions.

Removing the supply pipe. Cut a ½-inch section (see **Drawing 11**) out of the supply pipe near the elbow. If you're dealing with plastic or copper, use a pipe cutter; with galvanized pipe, you'll need a hacksaw. To free the pipe, detach the shank couplings.

Detaching the elbow from the stubout. For galvanized pipe, use the opposing force of two tape-wrapped wrenches (see **Drawing 12**) to loosen the connection. If you're working with copper tubing, unfasten any mechanical connections or use a torch to melt soldered joints (see "Removing copper pipe," page 53). Plastic pipe will have to be cut. Remove everything attached to the stubout.

Installing the shutoff valve. Clean and prepare the exposed end of the stubout to accept the appropriate fitting. If the stubout isn't threaded, solder or cement a fitting to it. Screw the shutoff valve to the fitting or pipe (see **Drawing 13**) after applying pipe joint compound to the threads. Hand tighten the connection, then use tape-wrapped wrenches to snug it up. Try to line up the valve outlet directly below the fixture inlet.

Making the hookup. Cut and bend the flexible tubing to fit its head into the fixture inlet and its bottom end inside the shutoff valve outlet (see **Drawing 14**). Fasten the coupling to the faucet inlet shank with a basin wrench. Secure the compression fitting to the valve with an adjustable wrench.

Adding a second sink

Fortunately, if you decide to add a second sink next to an existing one, you can make the connections for the side-by-side sinks without cutting into the wall. The new sink can share the trap and drainpipe of the existing sink (see **Drawing A**), but you'll need to extend the hot and cold supply pipes.

CAUTION: Before doing any work, turn off the water at the fixture shutoff valves or main shutoff valve (see page 23). Open the faucet.

Only a few steps are involved in this installation. First, locate the new sink so its drain hole is no more than 30 inches away from and no more than 6 inches higher than the existing fixture's drain hole. But be sure to slope the drainpipe (¼ inch per foot).

Remove the tailpiece from the drain of the existing sink and install a slip-joint tee fitting above its trap. Run drainpipe from the slip-joint tee to a 90-degree slip-joint elbow on the new sink's tailpiece.

Remove the existing shutoff valves and install tee fittings behind them. Run new supply pipes (see page 64) from the tee fittings to the shutoff valves of the new sink. Then install the original shutoff valves as described above.

The new fixture installation must meet the following requirements.

- *The new drainpipe* must enter the existing pipe at a point low enough to make waste flow downhill.
- *The slope of the drainpipe* should not exceed ¼ inch for every horizontal foot of pipe; if it does, water will be sucked from the trap.
- *A separate vent* may be required for the new fixture (see page 61).

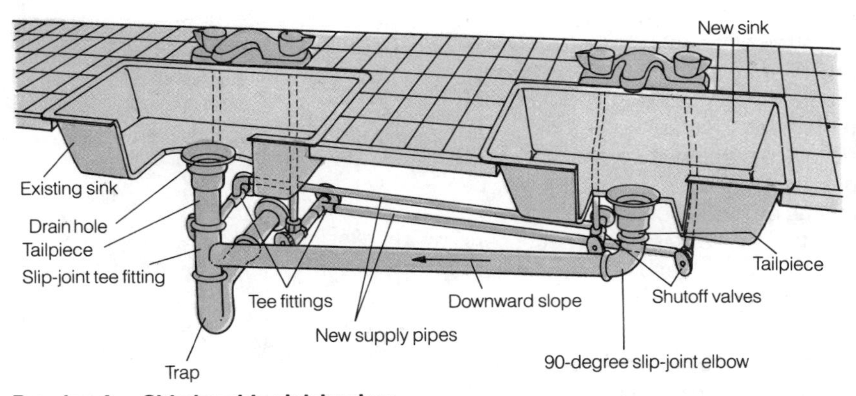

Drawing A. Side-by-side sink hookup

Replacing a Wall-hung Sink

Time to replace a damaged, outdated, or lackluster sink? You'll find the task is easier than you think.

Selecting a sink. When you shop for a new sink—whether it be a wall-hung or deck-mount style (see facing page)—you'll be faced with a choice of materials and configurations. Materials include stainless or enameled steel, porcelain-coated cast iron, plastic, and vitreous china. Configurations include single and double-well sinks in varying depths and shapes.

All sinks come with holes for either 4, 6, or 8-inch faucet assemblies. In most cases, faucets and sink flanges should be hooked up before the sink is installed.

Wall-hung sinks are not being used in residential plumbing as often as they used to be, but they can still be found in older homes. If you want to change from a wall-hung style to a deck-mount, see the installation instructions on the facing page. But if you need to replace a wall-hung with a new wall-hung model, the following guidelines will prove helpful.

CAUTION: Before doing any work, turn off the water at the fixture shutoff valves or main shutoff valve (see page 23).Open the faucet to drain the pipes.

Removing a wall-hung sink. Detach the faucet as described on pages 67–68. Disconnect the supply pipe couplings at the faucet inlet shanks under the sink to drain any remaining water from the pipes. Remove the trap from the tailpiece (see pages 24–25). Pull straight up on the sink (see **Drawing 15**) and it should lift off its hanger, brackets, or pedestal; if not, check underneath the sink for hold-down bolts that fasten it to the wall. The pedestal that a wall-hung sink may be resting on is often bolted to, or grouted into, the floor. Unfasten it, if necessary, and rock it back and forth to detach it from the floor.

Installing a bracing board. For a first-time sink installation or for a wall-hung sink that requires additional support, you'll need to recess a 2 by 6 or 2 by 8 bracing board (see **Drawing 16**) between two wall studs directly behind the sink. Notch the studs (see page 64) and nail or screw the bracing board into place; then finish the wall. You'll attach the sink hangers or brackets through the wall's surface.

Attaching hanging devices. Wall-hung sinks come with supporting hangers or angle brackets. Old-style pedestal sinks have special clips that screw to the wall. Refer to the manufacturer's instructions to properly position the hanging device on the wall.

Generally, the device will be centered, then leveled, over the drainpipe (see **Drawing 17**) at the desired height (31 to 38 inches above the floor). Fasten the hanging device to the wall with 3-inch woodscrews.

CAUTION: Angle brackets can give way under downward pressure. Install front supporting legs on the sink for extra stability.

Hooking up the sink. Attach the faucet (see pages 67–69) and sink flange (see facing page). Carefully lower the sink onto the hanger. Some hangers have projecting tabs that fit into slots under the sink's back edge. Angle brackets bolt up into the sink's base.

If the fixture needs supporting legs (see **Drawing 18**), fasten them to the sink, then screw the adjusting section of each leg downward until the sink is level. Seal the sink-wall joint with a bead of caulking compound. Hook up the water supply pipes and trap, turn the water back on, and tighten any leaky connections.

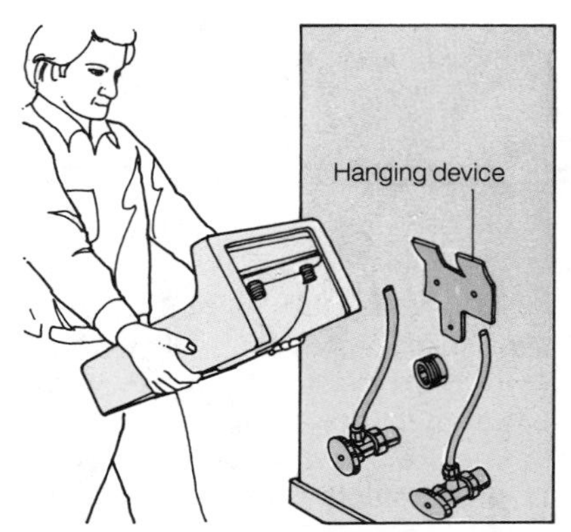

Drawing 15. Lift the sink off the wall-hanging device by pulling straight up on the fixture.

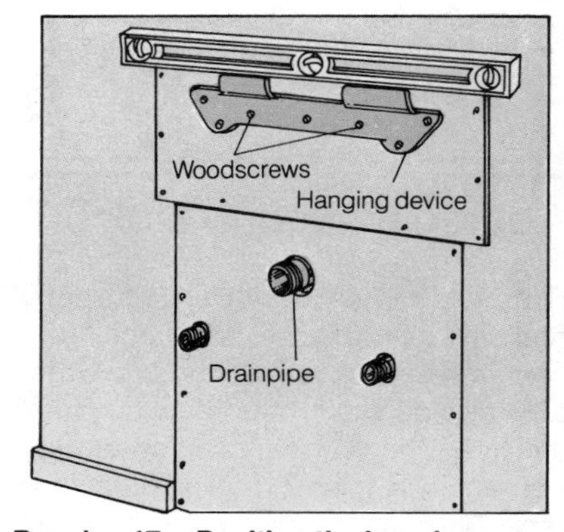

Drawing 17. Position the hanging device on the bracing board, centered and leveled directly over the drainpipe.

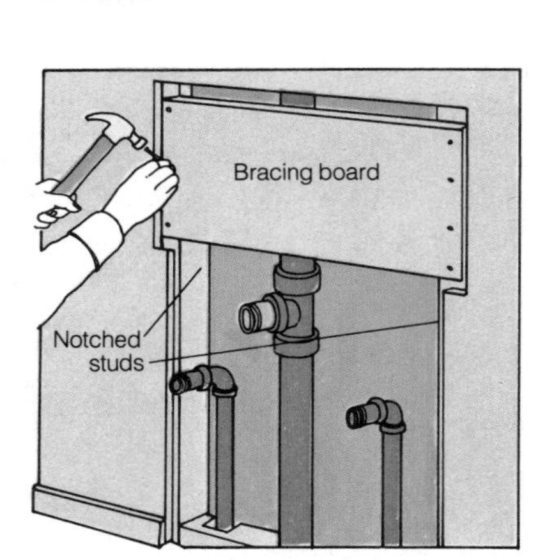

Drawing 16. Nail the bracing board between two notched studs in the wall in the spot where you plan to mount the sink.

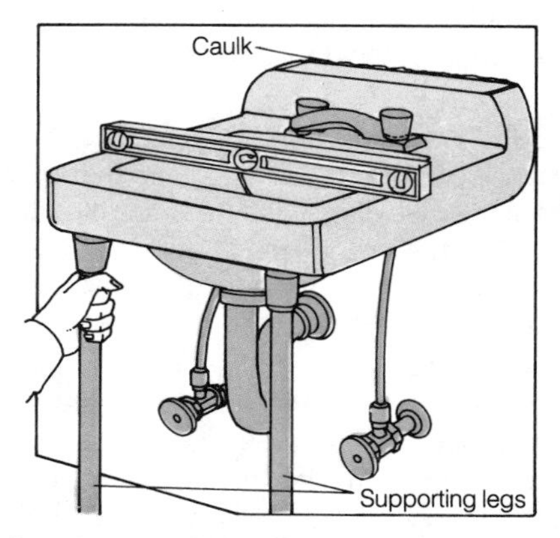

Drawing 18. Adjust the supporting legs until the sink is level, then run a bead of caulk along the sink-wall joint.

Installing a Deck-mount Sink

A close cousin to the wall-hung style, a deck-mount sink fits into a specially cut hole in a bathroom vanity or kitchen countertop. Whether frame-rimmed, self-rimmed, or unrimmed, a deck-mount sink (see **Drawing 19**) is sealed onto the countertop with clamps or lugs and plumber's putty. A frame-rimmed sink has a surrounding metal strip that holds the sink to the countertop. A self-rimmed sink has a molded overlap that's supported by the edge of the countertop cutout. Unrimmed sinks, though, are recessed beneath the countertop opening and held in place by metal clips.

If you're replacing a deck-mount sink, be sure to measure the hole in the countertop and take the measurements with you when you shop.

An increasingly popular variation of the deck-mount sink is the one-piece molded sink with an integral countertop. A time-saver to install, this type is simply set atop a cabinet and fastened from below.

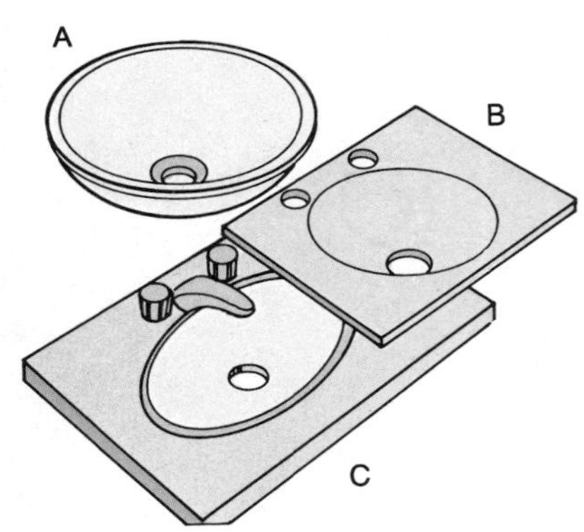

Drawing 19. Deck-mount sink styles: self-rimmed (A), unrimmed (B), frame-rimmed (C).

Removing the old sink. Begin by shutting off the water supply (see page 23), then draining and disconnecting the supply pipes (pages 70–71) and trap (pages 24–25). Remove a self-rimmed sink by forcing it free from below. With frame-rimmed and unrimmed sinks, remove the clamps or lugs from below. Caution: Suspend the weight of the sink from above or find a helper to support it while you remove the last of the lugs. It should then lift right out.

Positioning the sink. For a new installation, trace a template (see **Drawing 20A**) or the bottom edge of the frame (see **Drawing 20B**) onto the exact spot where the sink will sit. Use a saber saw to cut out the countertop opening. It's best to mount the faucet (see pages 67–69) and hook up the sink flange before installing the sink in the countertop.

Installing a sink flange. Kitchen sinks usually have strainers in their drains (see page 20); bathroom sinks have pop-ups (page 21) and flanges. To install a flange, run a ring of plumber's putty around the water outlet. Press the flange into the puttied outlet (see **Drawing 21**) and attach the gasket, locknut, and drain body to the bottom of the flange. Screw the tailpiece onto the drain body.

Framing the sink. For a frame-rimmed sink, apply a ring of plumber's putty around the top edge of the sink. Fasten the frame to the sink, following the manufacturer's directions—some frames attach with metal corner clamps or lugs (see **Drawing 22A**), others with metal extension tabs that bend around the sink lip (see **Drawing 22B**). Wipe off excess putty.

Securing the sink. Before installing a deck-mount sink of any style, apply a ½-inch-wide strip of putty or silicone adhesive along the edge of the countertop opening. Set the sink into the hole, pressing it down. Smooth excess putty. Anchor the sink at 6 to 8-inch intervals (see **Drawing 23**), using any clamps or lugs provided. Hook up the supply pipes and trap. Turn on the water and check for leaks.

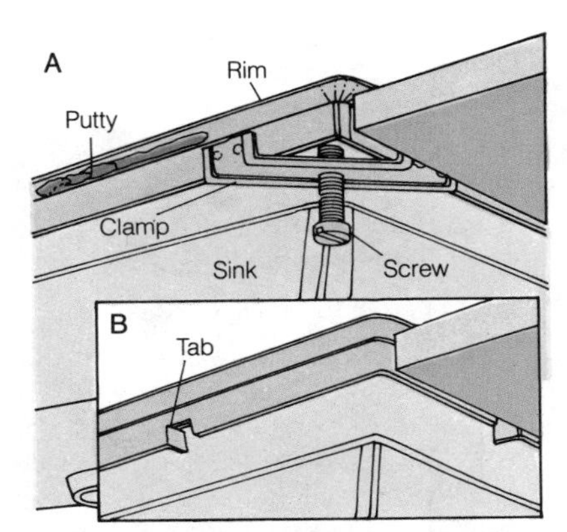

Drawing 22. Fasten the framing rim to the sink lip with either metal corner clamps or lugs (A) or bendable metal tabs (B).

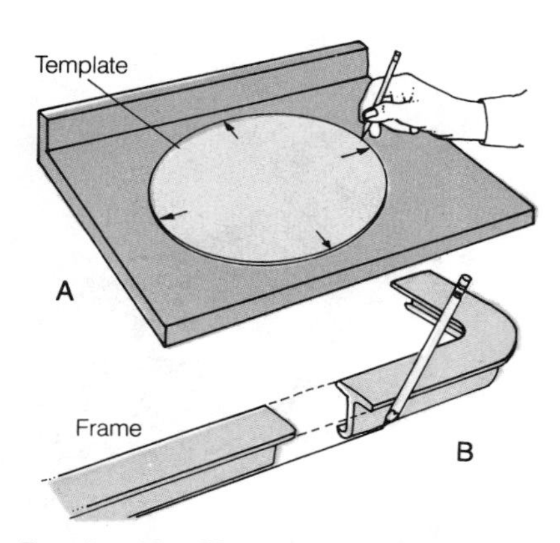

Drawing 20. Mark the exact spot where the sink will sit, by tracing either a template (A) or sink frame (B) onto the countertop.

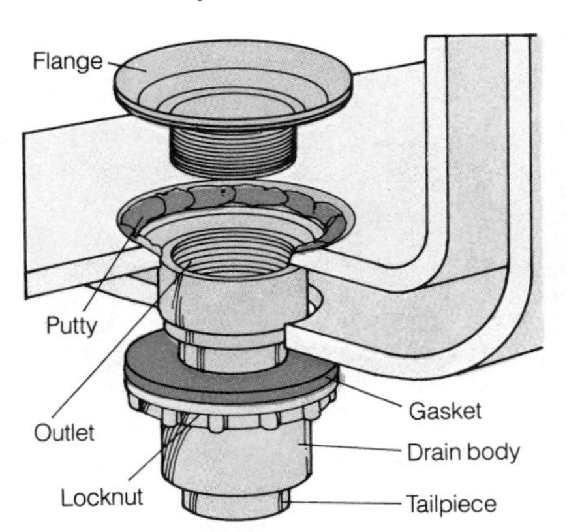

Drawing 21. Press the flange down into the puttied water outlet hole and attach the connecting parts from below.

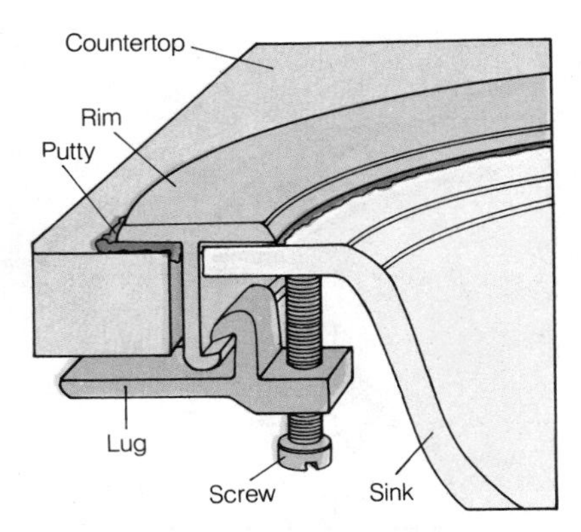

Drawing 23. Anchor the sink from below, using clamps or lugs every 6 to 8 inches, as needed.

Replacing Tub Faucets, Diverters & Shower Heads

Like sink faucets, tub faucets can be compression-style (see **Drawing 24**) or washerless (see **Drawing 25**). Either way, water is directed from the faucet to the tub or shower head by a diverter valve.

CAUTION: Before doing any work, turn off the water at the main shutoff valve (see page 23). Open the faucet to drain the pipes.

Disassembling a tub faucet. To disassemble the faucet, refer to pages 13–18. It's an easy job except for getting out the packing nut in a compression faucet. To get at the nut, chip away the wall's surface and grip the nut with a deep socket wrench (see **Drawing 26**).

Replacing a tub spout diverter. This diverter is in the tub spout and controls the water flow when the knob is lifted.

Grip the old spout with a tape-wrapped pipe wrench (see **Drawing 27**) and turn counterclockwise. Hand tighten the new spout into place.

Working on a shower head. Before replacing a leaking or water-wasting shower head, tighten all connections; if that doesn't stop the leak, replace the washer between the shower head and swivel ball (see **Drawing 28**).

If sluggish water flow is the problem, there's likely to be a blockage in the screen or face plate of the shower head. To cure this, remove the parts and clean them with a toothbrush.

Installing a shower head. Like a tub spout, a shower head simply hand screws onto the shower arm stubout. Before installing a new shower head, clean the pipe threads and apply pipe joint compound to prevent leaks.

Drawing 26. Get at the packing nut by chipping away the adjacent wall until you can turn the nut with a deep socket wrench.

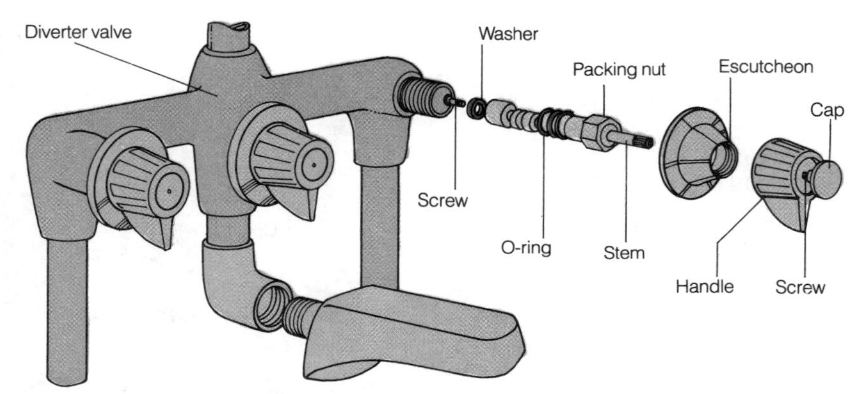

Drawing 24. Compression tub faucet

Drawing 27. Remove the old tub spout diverter by gripping the spout with a pipe wrench and turning counterclockwise.

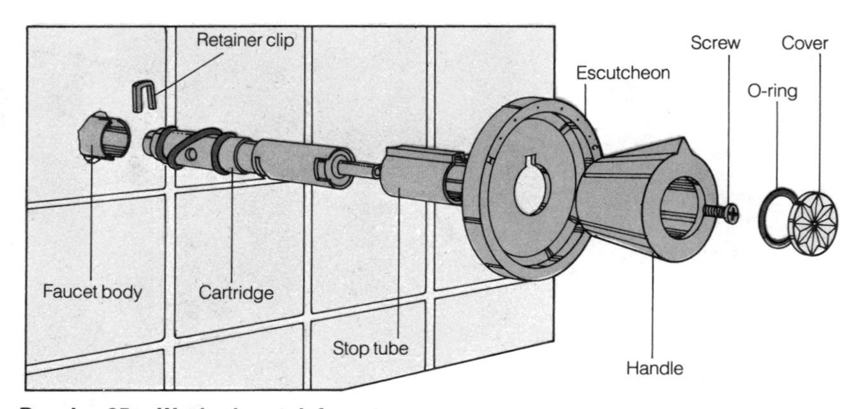

Drawing 25. Washerless tub faucet

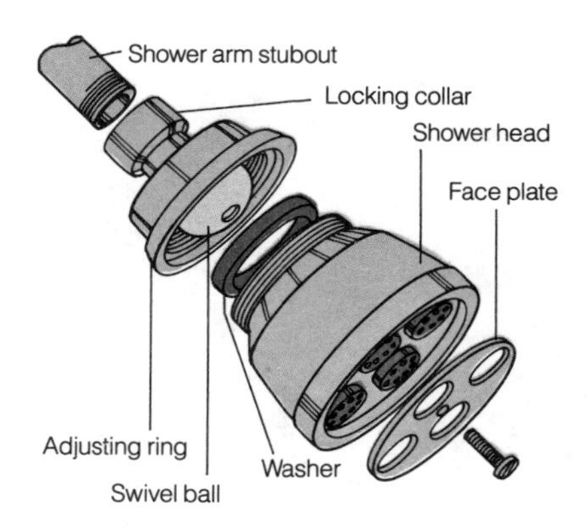

Drawing 28. To troubleshoot a leaking shower head, check the washer near the swivel ball for wear.

Hooking Up a Hand-held Shower

A hand-held shower offers an easy way to add or replace an over-the-tub shower, and it lets you direct the water as you please. One type of hand-held shower attaches directly to a shower-head diverter; another is an integral part of a tub spout.

To prevent tub water from siphoning back through a hand-held shower into the water supply, codes may require the installation of a vacuum breaker assembly. Manufacturers typically include instructions for installation of this device. Check your local code to see if one of them is required.

CAUTION: Before you begin work, turn off the water supply at the fixture shutoff valves or the main shutoff valve (see page 23). Open the faucet to drain the pipes of as much water as possible.

Attaching to a shower head. To begin, remove the existing shower head (see page 74) with tape-wrapped pliers to prevent marring the collar that secures the shower head. Clean the shower-arm threads with a wire brush and apply pipe joint compound.

Thread the diverter valve onto the shower arm. Use two wrenches to tighten (see **Drawing 29A**) until the diverter button is facing up. Fasten the hand-held shower hose to the shower head outlet on the diverter. Then screw the old shower head onto the shower hose outlet. Screw the hand-held spray head to the hose (see **Drawing 29B**). Turn on the water.

Attaching to a tub spout. For a hand-held shower that hooks up to a tub spout, you'll need to replace the spout with one that has a diverter valve knob and built-in hose outlet (see **Drawing 30**).

Begin by replacing the existing spout, as described on page 74. Apply pipe joint compound to the shower-hose outlet threads before attaching the hose. Then simply connect the hand-held spray head to the hose. To operate the diverter, lift the diverter knob on the spout and direct the spray head.

Mounting a hanger bracket. Most hand-held shower kits come with some type of wall bracket for hanging the unit. Position the hanger bracket on the wall at a convenient height. Mark and drill screw holes (see **Drawing 31A**) into the wall; if there's no stud to accept screws, use toggle bolts (hollow wall anchors) to fasten the bracket (see **Drawing 31B**).

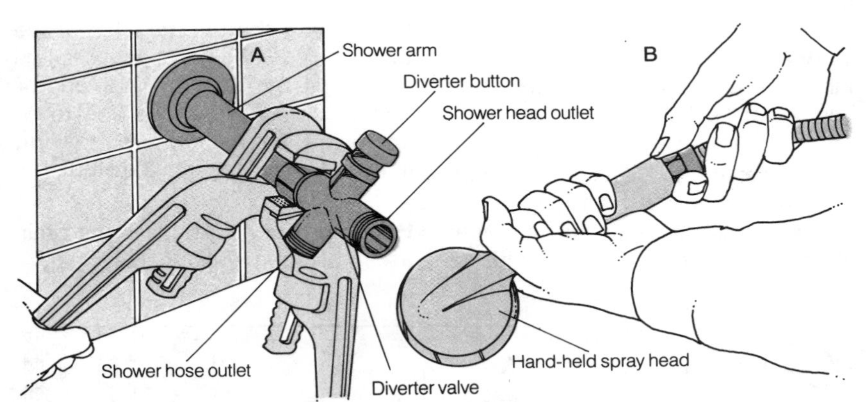

Drawing 29. Tighten the diverter valve onto the shower arm (A). Attach the shower hose, then connect the hand-held spray head to the shower hose (B).

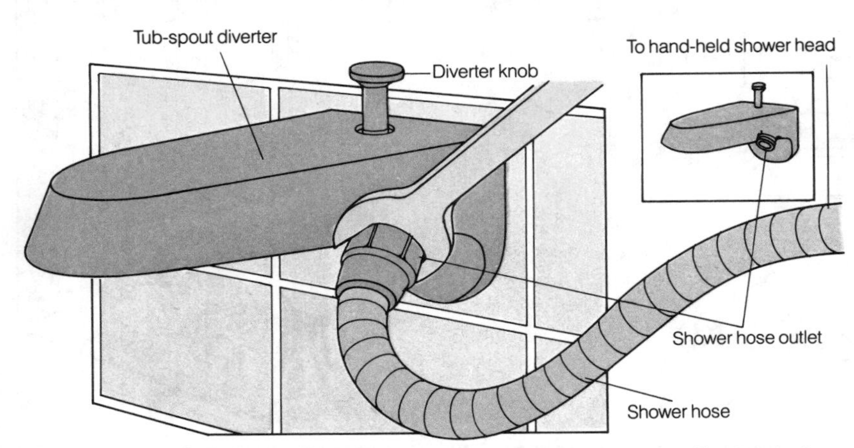

Drawing 30. Install a tub spout diverter in place of an existing spout to direct water to the hand-held shower head.

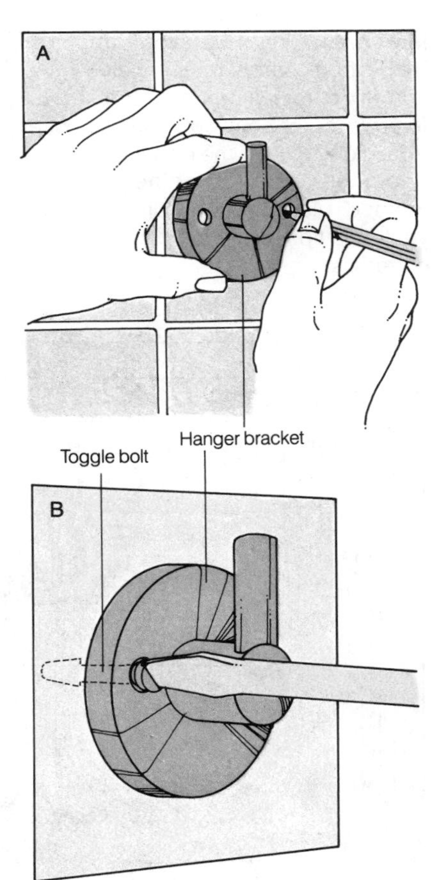

Drawing 31. Mark the spot where the hanger bracket will be (A) and mount it, using screws or toggle bolts (B).

Putting in a Tub & Shower

Whether you're replacing old fixtures or starting from scratch, tub and shower installations are complicated jobs. They take careful planning, precise carpentry, and a good bit of time. But the advantages are worth it, especially if you're adding an extra bathing spot that will put an end to early-morning traffic jams in the bathroom.

In plumbing supply stores you'll find a wide variety of sizes, shapes, and materials to tickle your decorating fancy.

Sizes & shapes. Before you go shopping, know the size of your existing fixture, if replacement is the issue. A standard tub is 5 feet long by 30 inches wide.

Shower stalls also come in various sizes and shapes, but should measure at least 32 inches across for comfort.

Materials. For tubs, your choices range from porcelain-enameled cast iron, to porcelain-enameled steel, to fiberglass; for showers, you'll choose from light metal, molded fiberglass, and tiled walls combined with a molded plastic base. On page 69 you'll find tips on caring for the various materials.

Removing a built-in bathtub

This is a major chore that requires lots of muscle and a good deal of demolition—stripping away walls and floor.

CAUTION: Before you do any work, turn off the water at the fixture shutoff valves or the main shutoff valve (see page 23). Open the faucet to drain the pipes.

To begin, remove the overflow plate, pop-up or strainer (see pages 20–21), and spout (see page 74) from the tub. Disconnect the supply pipes and drainpipes.

Chipping away tiles. If there are tiles along the wall or floor that come right up to the tub, chip them out. Remove the tile and plaster or wallboard from the enclosing walls (see **Drawing 32**), exposing several inches of the studs around the tub.

Getting the tub out. If yours is a steel tub, there may be nails or screws at the very top of the flange (lip), holding it in place. Remove them.

A cast iron tub is very heavy, but can be broken up with a sledge hammer; a steel tub is lighter, but it must be lifted out in one piece. If you decide to move a cast iron tub intact, plan to have at least four people to lift it. You may have to remove doors and even some trim to get any tub out.

Installing a new tub

A bathtub filled with water is extremely heavy. Check local building codes, and consider getting professional help for framing and support before you install a new tub.

Around its upper edge, your new tub will have a continuous flange that fits against the wall studs. There will also be two precut holes: one for the drain, the other for overflow. The faucet, spout, and shower head will mount on the wall above the tub. Piping for all of these must be in place and connected before you install the tub (see page 60).

Positioning the bathtub. Lower the tub into place so its continuous flange rests on 1 by 4 or 2 by 4 supports. Anchor the tub (except cast iron) to the enclosure by nailing or screwing through the flanges into the studs.

Hooking it up. Next, make the drain assembly connections (see pages

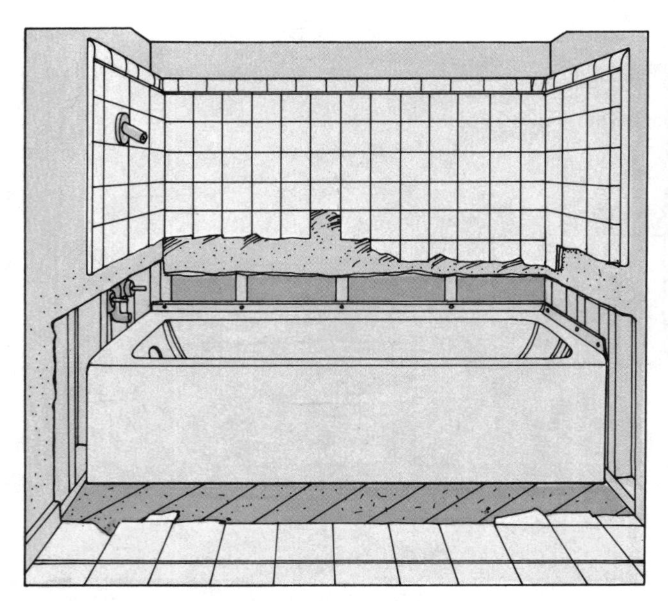

Drawing 32. To free a bathtub, tear away adjacent surface materials, such as tile, from the floor and enclosing walls, down to subfloor and wall studs.

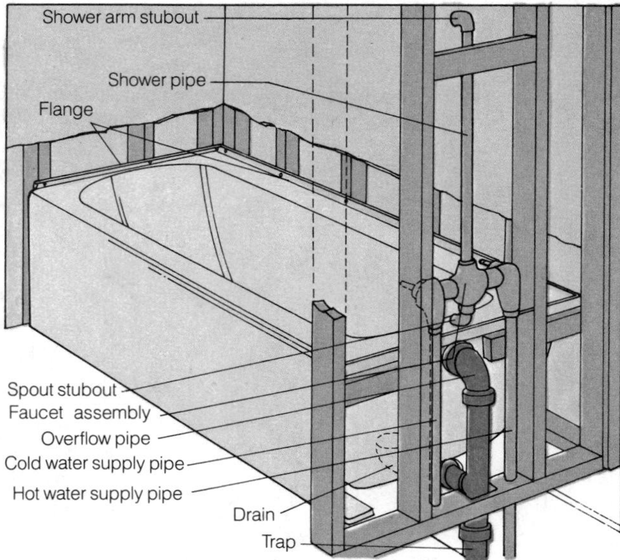

Drawing 33. When plumbing the bathtub, be sure to connect the overflow pipe with the drain above the trap.

24–25 and 29) through an access door in an adjoining room or hallway, or through an opening from below. The tub overflow must connect with the tub drain above the trap, not beyond it (see **Drawing 33**). Be careful not to overtighten the nuts.

Before surfacing the wall, restore the water pressure and check the drain and supply pipes for leaks. Use moisture-resistant wallboard as a base for the final waterproof wall covering; seal all joints between the tub and wall with silicone caulk to guard against water seepage. Complete the project by installing spout, faucet handles, and shower head.

Shower stall styles

When shopping for a shower stall, consider the following options.

Lightweight fiberglass units (see **Drawing 34A**), usually called shower wall surrounds, are the easiest to work with; they can be quickly set in place and attached to framing members. They're also easy to keep clean, thanks to their smooth surface and rounded corners.

Metal shower stalls (see **Drawing 34B**), quickly fading from the plumbing scene, are made of either tin or stainless steel. Tin is inexpensive but extremely noisy because of vibration. Stainless steel showers are more expensive and better looking, but they show water spots.

Tiled-wall showers (see **Drawing 34C**) have a molded shower base that sits on a waterproof foundation. The ceramic tile can be either individually installed tiles or pregrouted panels.

Adding a shower stall

First construct a frame to contain the shower unit. Be especially careful to make all measurements accurately and the framing square and plumb. Run supply pipes and drainpipes to the desired locations and install faucet, spout, shower head, and drain.

Positioning the shower stall. Set the shower stall in place. Be sure to read the manufacturer's installation instructions, since stalls vary. Attach the shower stall as you would a tub, nailing the flange to the studs.

Hooking it up. Drill holes into the surface of the stall for the shower valve assembly and shower arm stubout (see **Drawing 35**). Again, refer to the installation instructions for exact locations and how-to information.

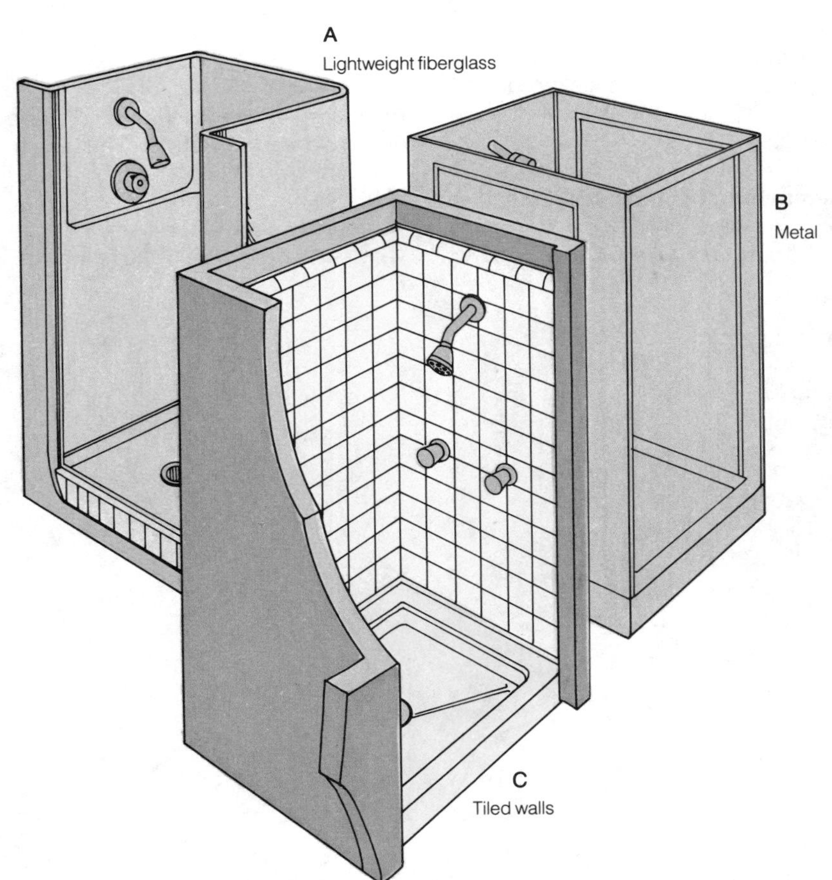

Drawing 34. Types of shower stalls include lightweight fiberglass (A), tin or stainless steel (B), and tiled-wall units (C).

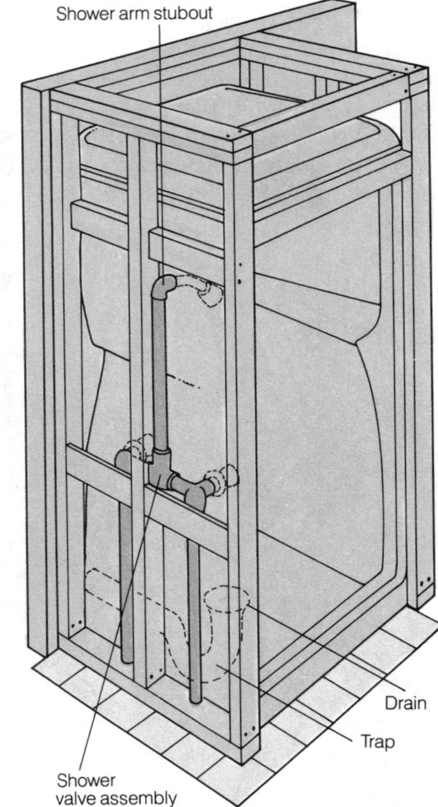

Drawing 35. When plumbing a shower stall, run supply pipes for shower valve assembly and shower arm stubout.

Replacing a Toilet

If your toilet has seen better days, you'll be glad to know that replacing it is a one-afternoon project that you can tackle yourself. Installing a toilet in a new location is more of a challenge, because of the need to extend supply and drainpipes (see pages 60–65). You may want to have a professional run the pipes to the desired spot, then do the installation yourself.

Making a selection. When shopping for a toilet, you'll find lots of choices —wall-mount, bowl-mount, water-saver, wash-down, reverse-trap, and siphon-jet models. These are interchangeable.

Determining roughing-in size. Other than code requirements for a new toilet, the only crucial dimension to consider when installing a toilet is its roughing-in size—the distance from the wall to the center of the drainpipe.

You can usually determine roughing-in size without removing the bowl—just measure from the wall to the center of the two hold-down bolts that secure the bowl to the floor. (If the bowl has four hold-down bolts, measure to the rear pair.)

The roughing-in distance for the new toilet can be shorter than that of the old fixture, but it cannot be longer or the new toilet won't fit.

Buying parts. Pick out a model that's ready to install—one that has a flush mechanism already in the tank. With the toilet, you'll get the necessary gaskets, washers, and hardware for fitting the tank to the bowl, but you may need to buy hold-down bolts and a wax gasket.

Also, buy a can of plumber's putty to secure the toilet base to the floor and the caps to the hold-down bolts. Finally, if the old toilet didn't have a shutoff valve, consider installing one (see pages 70–71).

Disconnecting the water supply. First, turn off the water at the fixture shutoff valve or main shutoff valve (see page 23). Flush the toilet twice to empty the bowl and tank. Sponge out any water that remains. Unfasten the coupling on the flexible tubing (see **Drawing 36**) at the bottom of the tank. If the flexible tubing is kinked or corroded, replace it with new flexible tubing.

Removing the tank. If yours is a bowl-mounted toilet, unbolt the empty tank from the bowl (see **Drawing 37A**), using a screwdriver to hold the mounting bolt inside the tank while unfastening its nut with a wrench from below.

If the tank is wall mounted (see **Drawing 37B**), use a spud wrench to loosen the couplings on the pipe connecting the tank and bowl. Remove the pipe and unscrew the hanger bolts that attach the tank to the wall. If the replacement toilet is not wall mounted, remove hanger brackets.

Removing the bowl. Pry the caps off the hold-down bolts and remove the nuts with an adjustable wrench. If the nuts are rusted on the bolts, soak them with penetrating oil or cut the bolts off with a hacksaw.

Gently rock the bowl from side to side to break the seal between the bowl and the floor. Lift the bowl straight up (see **Drawing 38**), keeping it level. Stuff a rag into the drainpipe to minimize unpleasant sewage odors and to keep debris from falling into the opening.

Preparing the floor flange. Scrape up the old wax gasket with a putty knife (see **Drawing 39**) and remove the old hold-down bolts from the floor flange. Thoroughly scrape the flange to prevent leaks at the base of the new bowl.

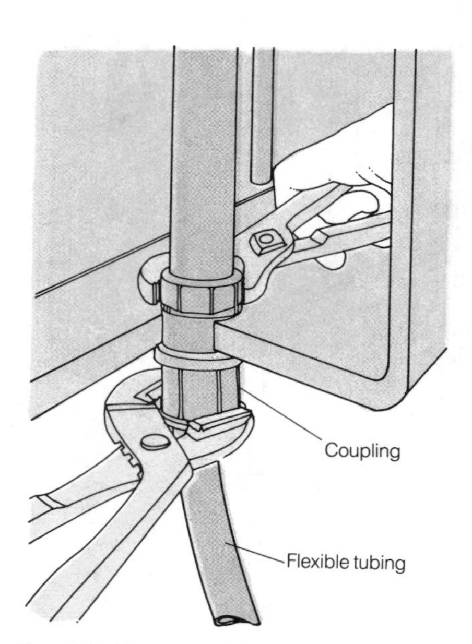

Drawing 36. Loosen the coupling on the flexible tubing at the bottom of the tank, using the force of two pliers.

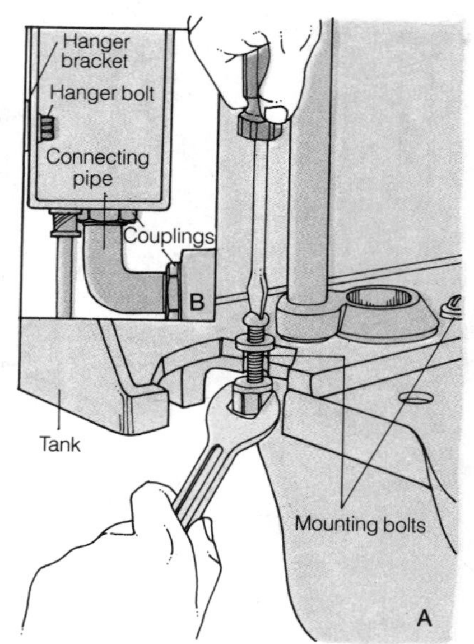

Drawing 37. Detach the tank from the bowl by loosening the mounting bolts (A) or by unfastening the couplings (B).

Drawing 38. Lift the bowl straight up off the floor flange, keeping it level to avoid spilling any remaining water.

Check the floor flange for deterioration. If it's cracked or broken—or if you just want to guard against trouble later—replace it with a copper or plastic one that can be soldered or cemented into place. Set the new floor bolts in plumber's putty and insert them through the flange. Adjust the bolts so that they line up with the center of the drainpipe.

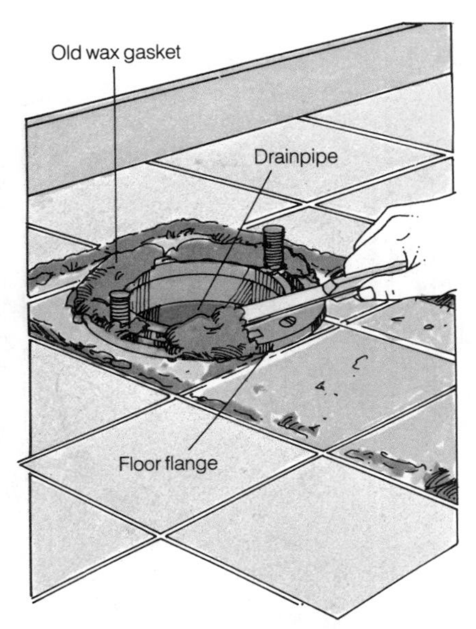

Drawing 39. Thoroughly scrape up the old gasket from the floor flange, using a putty knife or similar tool.

Installing the wax gasket. Turn the new bowl upside down on a cushioned surface. Place the new wax gasket (see **Drawing 40**) over the toilet horn (outlet) on the bottom of the bowl. Apply plumber's putty to the bottom edge of the bowl.

Placing the bowl. Remove the rags from the drainpipe. Gently lower the bowl into place atop the flange, using the bolts as guides. Press down firmly, while twisting slightly.

Checking with a level, straighten the bowl (see **Drawing 41**), and use thin pieces of metal to shim the bowl where necessary. Hand tighten the washers and nuts onto the bolts.

Attaching the tank. Fit the rubber gasket onto the flush-valve opening (see **Drawing 42**) on the bottom of the tank. Place the rubber tank cushion on the rear of the bowl. Position the tank over the bowl and tighten the nuts and washers onto the mounting bolts.

Use an adjustable wrench to snug up the hold-down nuts at the base of the bowl. Check that the bowl is still level. Fill the caps with plumber's putty and place them over the bolt ends. Use caulking compound to seal the base of the toilet bowl to the floor.

Hooking up the water supply. If your water supply stubout comes from the wall and the new tank is lower than the original, install an elbow fitting on the stubout. Use two 4 to 6-inch threaded nipples and a second elbow to connect the shutoff valve to the flexible tubing (see **Drawing 43**). Fasten the flexible tubing to the bottom of the tank.

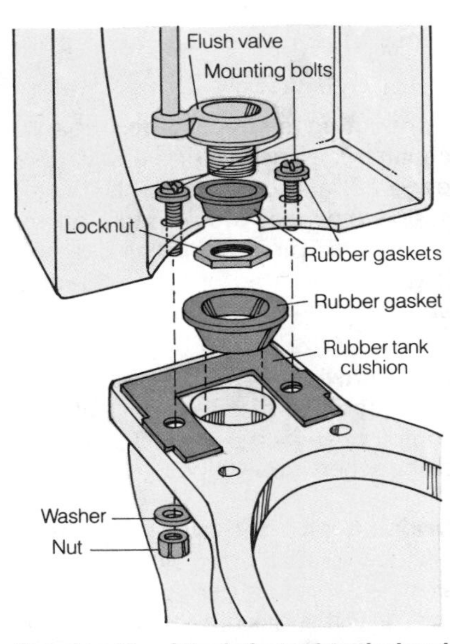

Drawing 42. Attach the tank to the bowl using the mounting bolts, with the rubber gasket and tank cushion in place.

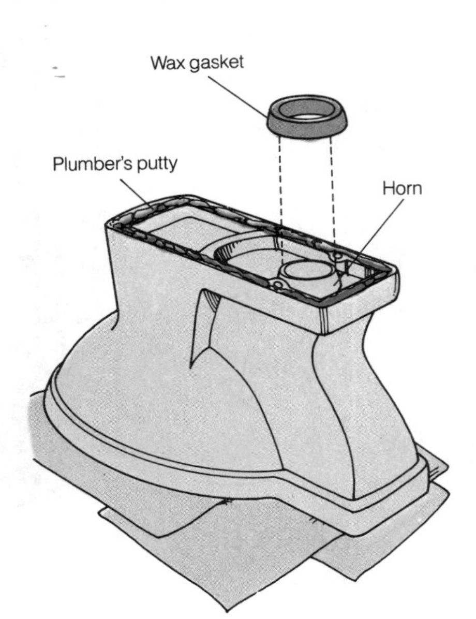

Drawing 40. Position the new wax gasket over the toilet horn on the bottom of the bowl.

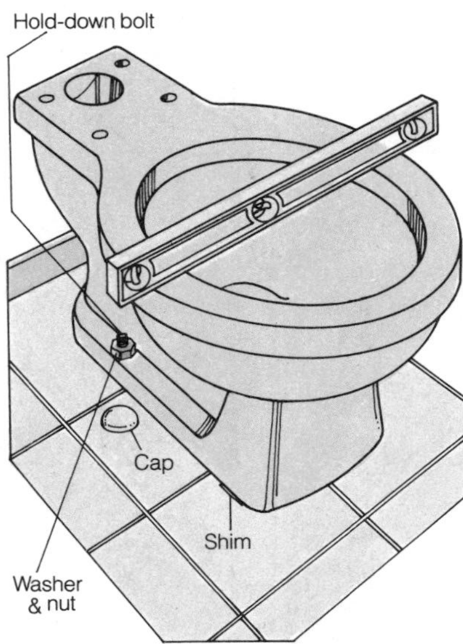

Drawing 41. Level the bowl once it's in place, using small pieces of metal to shim it, if necessary.

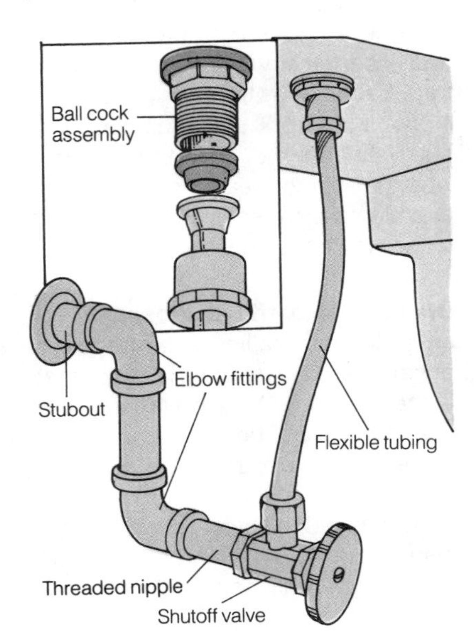

Drawing 43. Fasten the shutoff valve to the flexible tubing, using connecting elbows and nipples to adjust the pipe height.

Installing a Dishwasher

If you're ready for your first dishwasher or need to replace a well-used one, you can do the plumbing portion of the installation yourself. It's advisable, though, to call a professional to make the electrical connections.

A built-in dishwasher requires a hot water supply pipe connection (cold isn't necessary), a drainpipe fitting, and a venting hookup.

Some municipalities require a permit and an inspection when a built-in dishwasher is installed.

Connecting to supply pipe. Begin by shutting off the water supply (see page 23) and draining the hot water supply pipe that you plan to tap into. Cut into the pipe and install a tee fitting, as described on page 63, or a special three-way valve.

Run flexible tubing from the tee (see **Drawing 44**) to the water inlet valve for the dishwasher. Install a shutoff valve (see pages 70–71) in the dishwasher supply pipe.

Draining into the sink trap. Your dishwasher can drain either into the sink trap or into a garbage disposer. For use with a sink drain, you'll need to buy a threaded waste tee fitting (see **Drawing 45**).

To install the waste tee, remove the sink tailpiece (see pages 24–25) and insert the waste fitting into the trap. Secure it by tightening the coupling on the trap. Cut the tailpiece so it fits between the sink strainer and waste tee fitting. Reattach the tailpiece and clamp the dishwasher drain hose to the waste tee fitting.

Draining into the disposer. If you have, or are installing, a garbage disposer, the dishwasher drain hose attaches to it (see **Drawing 46**). To make the connection, begin by turning off the electrical circuit to the disposer. Use a screwdriver to punch out the knockout plug inside the disposer's dishwasher drain fitting. Clamp the dishwasher drain hose to the fitting.

Venting the dishwasher. Most codes require you to connect an air gap (see page 41) to the dishwasher's drain hose to prevent the siphoning of tainted water from the sink drain back into the dishwasher. Where this is not required, you can make a gradual loop with the drain hose to the height of the dishwasher's top instead.

Completing the installation. Slide the dishwasher into place. Make the supply and drain hookups (see **Drawing 47**) according to manufacturer's directions.

Once it's hooked up, level the dishwasher by adjusting the height of its legs. Anchor the unit to the underside of the counter with the screws provided. Finally, restore water pressure and check for leaks.

Drawing 44. Install a tee fitting in the water supply pipe and run flexible tubing from it to the dishwasher.

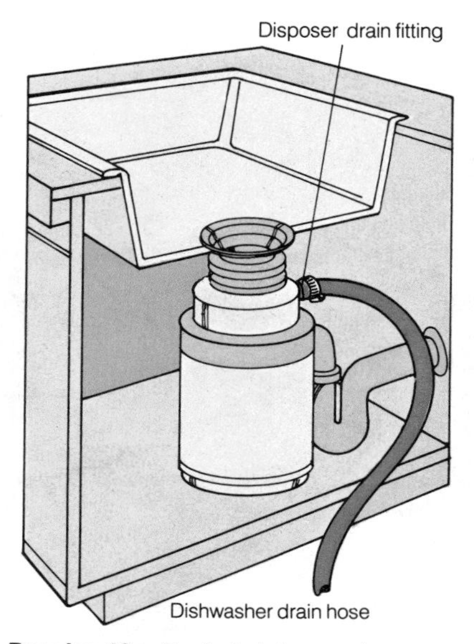

Drawing 46. To drain into a garbage disposer, connect the dishwasher drain hose to the disposer's drain fitting.

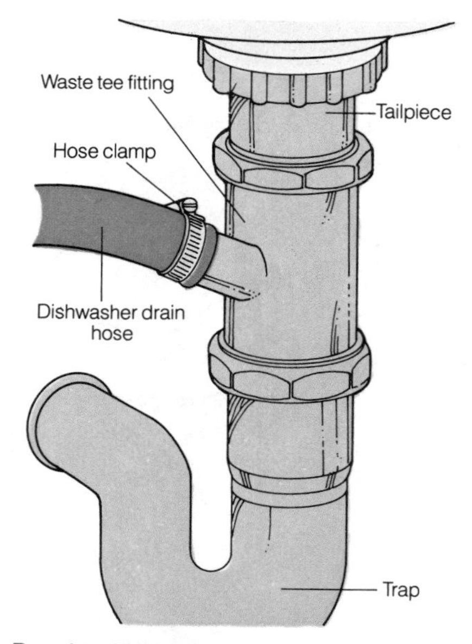

Drawing 45. Add a threaded waste tee fitting above the trap into which used water from the dishwasher will drain.

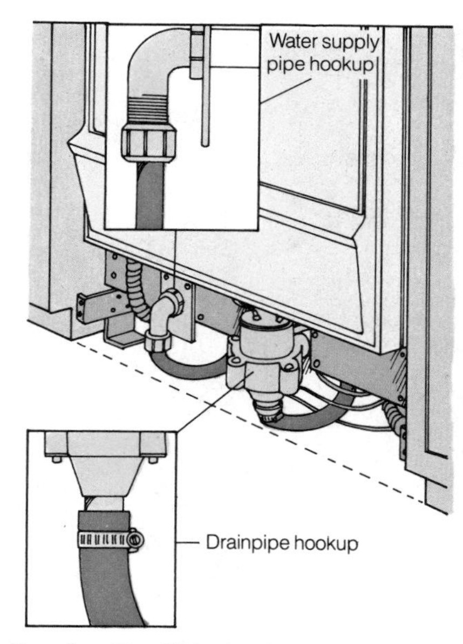

Drawing 47. Make the supply and drain hookups as specified in the manufacturer's instructions for the dishwasher model.

Hooking Up a Garbage Disposer

Garbage disposers have proven to be efficient labor-saving devices that gobble up almost anything you feed them. Disposers have become necessary kitchen tools—and with good reason. What better way to get rid of food waste quickly and easily? Today's disposers are quiet, efficient, and virtually trouble-free.

Before you install a disposer, check codes in your area for any restrictions. And for tips on fixing and caring for garbage disposers, see page 40.

Installation takes a few hours, but the work isn't very difficult. Most units fit the standard 3½ or 4-inch drain outlets of kitchen sinks and mount somewhat like a sink strainer. But like all products, disposers vary from brand to brand.

Plumbing a disposer involves altering the sink trap to fit the unit. Some models have direct wiring that should be connected by a licensed electrician. Other plug-in disposers require a 120-volt grounded outlet under the sink—another job for an expert in electrical work.

If you're replacing a disposer, turn off the electricity, then unplug the unit or disconnect the wiring before removing the disposer. Always be extremely cautious when working with a combination of plumbing and electricity—water and watts don't mix.

You'll get detailed installation instructions with your disposer, but here are the typical steps to follow when you hook one up.

Removing the sink strainer. Disconnect the tailpiece and trap (see pages 24–25) from the sink strainer. Disassemble the sink strainer (see page 20) and lift it out of the sink. Clean away any old putty or sealing gaskets from around the opening.

Installing the mounting assembly. The disposer will come with its own sink flange and mounting assembly. Run a rim of plumber's putty around the sink opening and seat the flange in it. Then, working from below, slip the gasket, mounting rings, and snap ring (see **Drawing 48**) up onto the neck of the sink flange. The snap ring should fit firmly into a groove on the disposer's sink flange to hold things in place temporarily.

Uniformly tighten the slotted screws in the mounting rings until the gasket fits snugly against the bottom of the flange. Remove any excess putty from around the flange.

Fastening the disposer. Attach the drain elbow to the disposer. Align the holes in the disposer's flange with the slotted screws in the mounting rings.

Rotate the disposer so that the drain elbow lines up with the drainpipe (see **Drawing 49**). Tighten the nuts securely onto the slotted screws to ensure a good seal.

Hooking up the drain. Fit one of the couplings and a washer onto the drain elbow (see **Drawing 50**), then fasten the trap to the drain elbow. Add an elbow fitting onto the other end of the trap to adapt to the drainpipe. You may need to cut the elbow to make the connection.

Run water down through the disposer to test for leaks. Tighten any loose connections.

Energizing the disposer. The last part of the installation requires electrical know-how. It's best to call in an electrician to run the wires from a power source to an outlet under the sink and to the disposer ON/OFF switch.

If you do the work yourself, be sure to shut off power before beginning. Also, check that the unit is properly grounded before restoring the power.

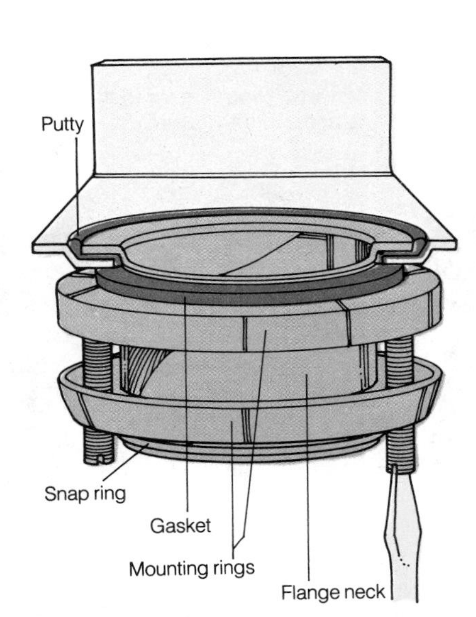

Drawing 48. Attach the mounting rings, with gasket and snap ring, to the sink flange and tighten the screws.

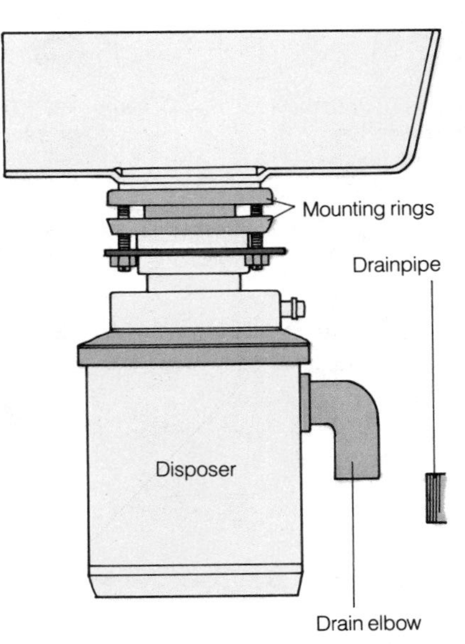

Drawing 49. Line up the drain elbow on the disposer so it is directly opposite the drainpipe.

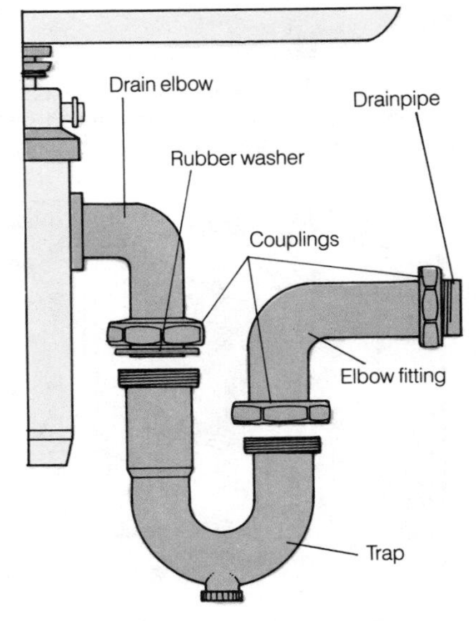

Drawing 50. Connect the trap to the disposer's drain elbow and to the elbow fitting on the drainpipe.

Adding a Hot Water Dispenser

Does a built-in appliance that delivers steaming hot water in an instant—for coffee, tea, or soup—sound appealing? A hot water dispenser that mounts on your kitchen sink or countertop eliminates the need to boil water each time you want something hot to drink, and it saves energy.

These easy-to-install dispensers have a stainless steel faucet that connects to an undercounter storage tank. The tank, which taps into the nearby cold water pipe, has an electric heating coil that keeps water at about 200°—about 50° hotter than that produced by the average water heater.

The unit must be connected to a 120-volt, three-prong outlet installed under the sink. Some models are directly wired to a grounded electrical receptacle, and most have a switch that lets you turn the tank heater off.

CAUTION: A hot water dispenser's spout, the water coming from it, and the storage tank are very hot.

Installing a hot water dispenser. Start by deciding where you want to put the unit. Commonly, its faucet fits in a hole at the rear of the sink rim or mounts directly on the countertop. For the latter, drill a 1¼-inch hole in your countertop near the cold water faucet. Following manufacturer's instructions, attach the dispenser faucet from beneath the sink (see **Drawing 51**). Generally, you'll need only to tighten a wing nut with washers to hold the faucet.

Mounting the tank. Screw the tank mounting bracket to the wall or cabinet back (see **Drawing 52**) making sure it's plumb. It should be located about 14 inches below the underside of the countertop. Next, mount the tank on it.

Installing a tee fitting. Before plumbing the unit, shut off the water (see page 23) and drain the pipes. Many dispensers come with a self-tapping valve. If yours doesn't, and if codes permit, tap into the cold water pipe using a saddle tee fitting (see **Drawing 53A**). To do this, clamp the fitting to the supply pipe, then drill a hole in the pipe.

If saddle tee fittings aren't permitted in your area, tap in with a standard tee fitting and install a shutoff valve and adapter (reducer) fitting (see **Drawing 53B**) for the dispenser.

Making the hookups. Using the compression nuts provided with the unit, attach one incoming water supply tube from the dispenser to the storage tank, and another from the tank to the cold water supply pipe (see **Drawing 54**). Turn on water; check connections.

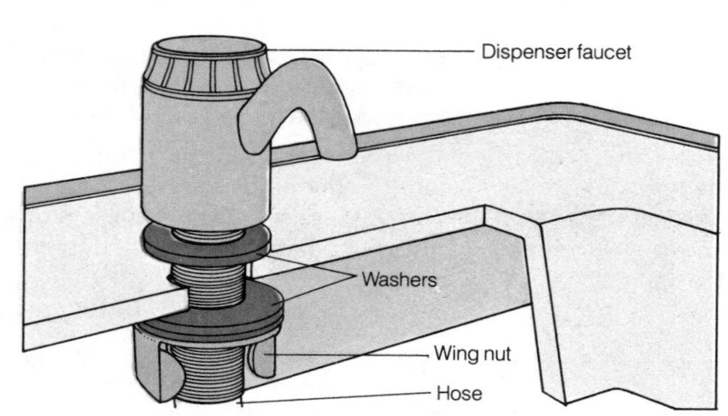

Drawing 51. Fasten the dispenser faucet from beneath the sink, tightening the wing nut and washers on the dispenser assembly until secure. Follow manufacturer's instructions.

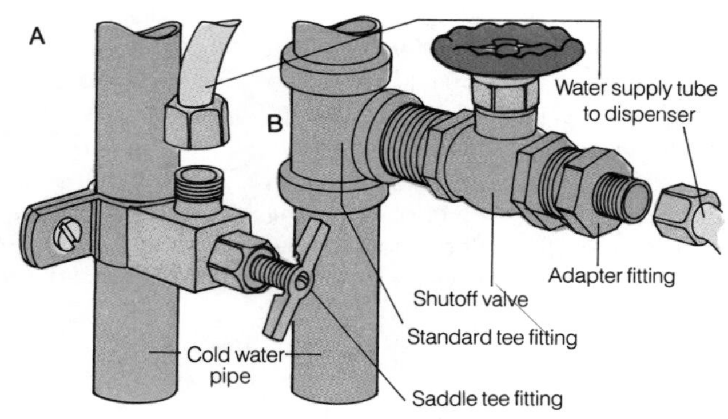

Drawing 53. Tap into the cold water pipe with a saddle tee fitting (A) or a standard tee fitting and shutoff valve (B).

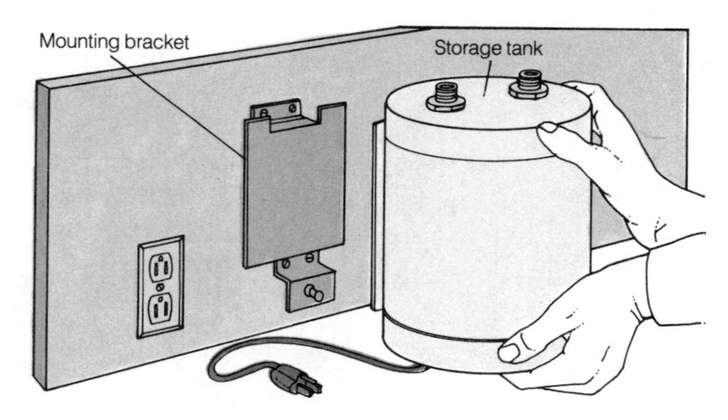

Drawing 52. Attach the tank mounting bracket to the wall or cabinet using woodscrews; check plumb. Install the storage tank on the mounting bracket.

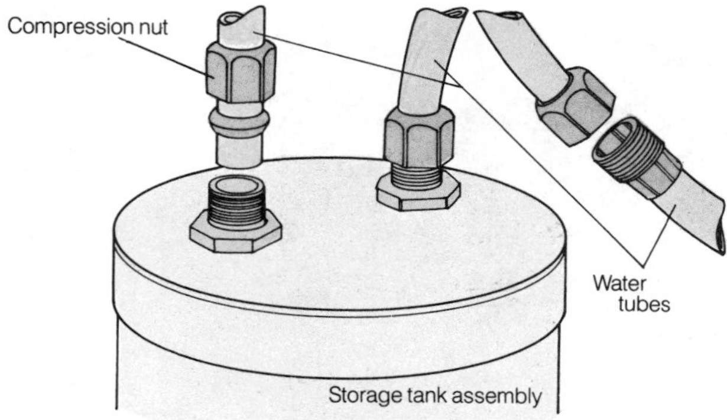

Drawing 54. Attach the water tubes between the storage tank and the dispenser, and between the tank and the supply pipe.

Connecting a Washing Machine

Adding the necessary plumbing connections for a washing machine is a fairly straightforward task. Most building codes now require that new residential units have the connecting pipes already installed, but if you live in a home that's never had a washer, you can do the plumbing yourself.

Readying supply pipes & drainpipes

You'll need to run both hot and cold water supply pipes to the desired location. In addition, each supply pipe needs a shutoff valve (see pages 70–71) and an air chamber (see page 45) to prevent banging.

To start, locate and drain the nearest hot and cold water pipes. Supply pipes for an automatic washer are usually ½ inch in diameter. Check your local code and also the manufacturer's instructions before you install supply pipes. Extend the pipes (see pages 60–64) to the desired point just above the washer and install a tee fitting at the end of each pipe.

If there's no sink or laundry tub nearby, you'll need to drain the washer into a special drainpipe called a standpipe—a 2-inch-diameter pipe with a built-in trap that taps into the nearest drainpipe.

Minimizing banging pipes

Attach air chambers (see **Drawing 55**) to the tee fittings in the hot and cold water pipes to minimize water hammer, which is noisy and destructive. Some manufacturers recommend that air chambers (see page 45) be of pipe that's one size larger than the supply pipes themselves and that the chambers be as long as 24 inches—1½ to 2 times longer than ordinary ones.

Choosing shutoff valves

Extend the pipes from the tee fittings, leaving enough space above the washing machine for shutoff valves. You can install either two threaded-spigot shutoff valves or a single-lever valve. Either way, it's a good idea to close the shutoff valves any time the washing machine is not in use. This relieves the constant water pressure on the supply hoses and the water inlet valve—and could prevent a flood.

Threaded-spigot shutoff valves (see **Drawing 56**) are used most often. To install, add elbows at the end of the supply pipes and attach threaded nipples, then threaded spigots to accept the machine hoses.

A single-lever shutoff valve (see **Drawing 57**) turns off both hot and cold water simultaneously with just a flick of the wrist. This valve can be installed in place of existing valves, with little or no modification.

Unscrew the valve adapters from the single-lever unit and solder (see pages 54–55) one to the end of each supply pipe. (Use transition fittings—see page 53—if your pipes aren't copper.) Slide the gaskets and valve body onto the valve adapters. Insert and tighten the attachment screws, then thread on the washer hoses.

Draining into a standpipe

The standpipe, available in lengths from 34 to 72 inches, should stand taller than the highest level of water in the washer to prevent backup and siphoning of dirty water into the machine. To determine the size standpipe you'll need, check the manufacturer's instructions.

To install a standpipe (see **Drawing 58**), cut into a drainpipe and install a sanitary tee fitting. Attach the standpipe to the tee and push the washing machine's drain hose down into the standpipe about 6 inches—be sure the hose won't be forced out of the pipe by the water pressure.

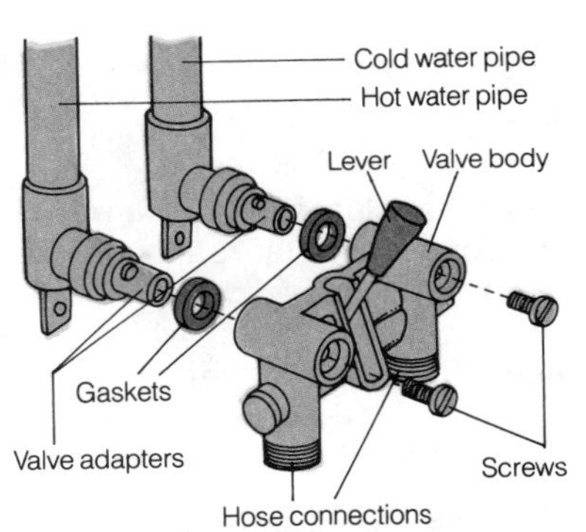

Drawing 57. Install a single-lever valve plus its adapters onto the ends of the hot and cold water pipes.

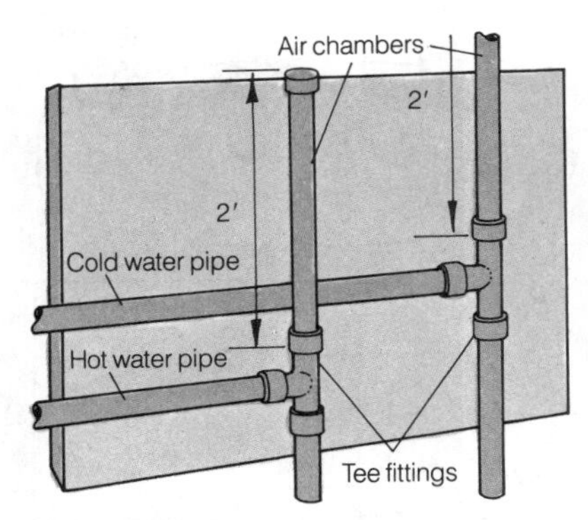

Drawing 55. Attach air chambers to tee fittings that are installed in the hot and cold water pipe extensions.

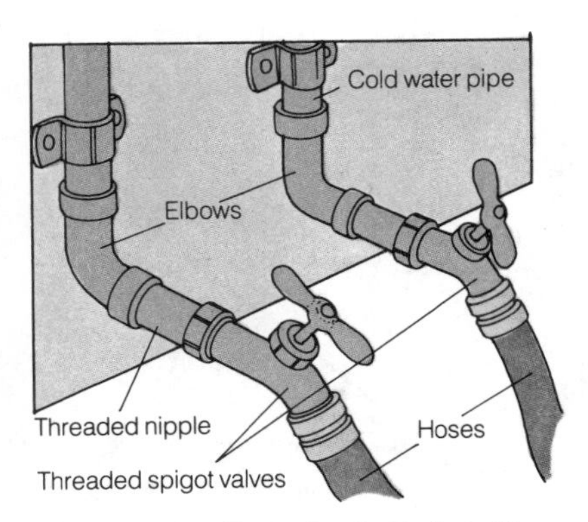

Drawing 56. Thread spigot shutoff valves onto elbows at the ends of both the hot and cold water pipes.

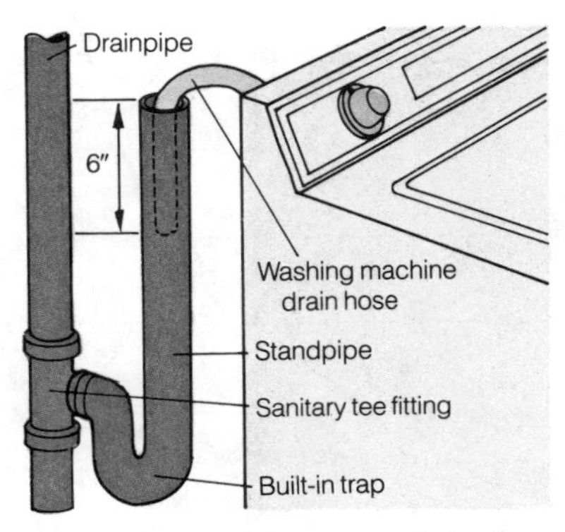

Drawing 58. Attach the standpipe to a sanitary tee fitting in the drainpipe and insert the washing machine drain hose.

Replacing a Water Heater

It's time to start budgeting for a new water heater when yours begins to leak or show rust and corrosion. When you go shopping for a new water heater, you'll need to consider four factors: capacity, warranty, tank lining, and recovery rate.

Capacity. Gas heaters are usually sized from 30 to 100 gallons; electric heaters hold up to 102 gallons. The graphs below will give you a rough idea of the size water heater tank you need. The capacity should be based on the number of people in the household and the number of bathrooms you have.

Warranty. Most water heaters come with a 7 to 15-year warranty. It usually pays to choose the top-of-the-line model, but beware: some models that manufacturers call deluxe energy-savers cost significantly more and carry a longer warranty on their tank, but they don't necessarily cost less to operate.

Lining. Glass, the most common lining for water heater tanks, lasts longer and provides cleaner water than other liners. Copper-lined tanks, though, are better and longer-lived than glass-lined galvanized tanks, which are usually the least expensive and shortest-lived, because of corrosion caused by chemicals in the water.

Recovery rate. This refers to the number of gallons per hour that a heater can raise to 100°. Generally, gas heaters have the fastest rate, electric heaters the slowest. Manufacturer's product literature will provide information on recovery rate.

Fuel. If you're replacing your old water heater, it's almost always preferable (and easier) to stay with one that uses the same type of fuel. In a new installation, availability and cost of gas versus electricity should be your primary consideration.

CAUTION: If you need to run gas piping to a new water heater, it's best to call a professional to do the work. If you're installing an electric water heater, use extreme caution when making the electrical hookup.

Emptying the tank. When you're ready to replace your water heater, start by shutting off the water and fuel (or power) supply to the old unit (see **Drawing 59**). If there's no floor drain beneath the valve, connect a hose to the valve and run it to a nearby drain or outdoors. Then drain all the water out of the heater storage tank by opening the drain valve (see drawings on page 38) near the base of the tank.

Disconnecting supply pipes. Next, disconnect the water inlet and outlet pipes from the heater. If they are joined by unions (see **Drawing 60A**) or flexible pipe connectors the job is simple—just unscrew them. If not, you'll have to cut through the pipes with a hacksaw (see **Drawing 60B**).

Detaching the power or fuel line. To disconnect the power supply lines on an electric heater, shut off the power; remove the electrical cable from the heater. For a gas or oil water heater, shut off the gas and use a wrench to

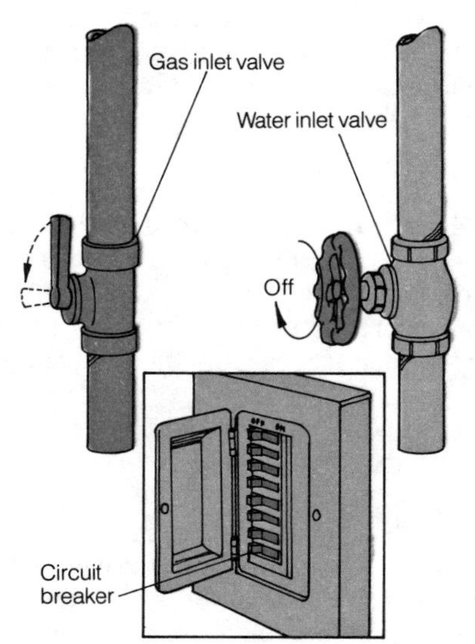

Drawing 59. Shut off the fuel (or power) and water supply to the water heater before doing any work.

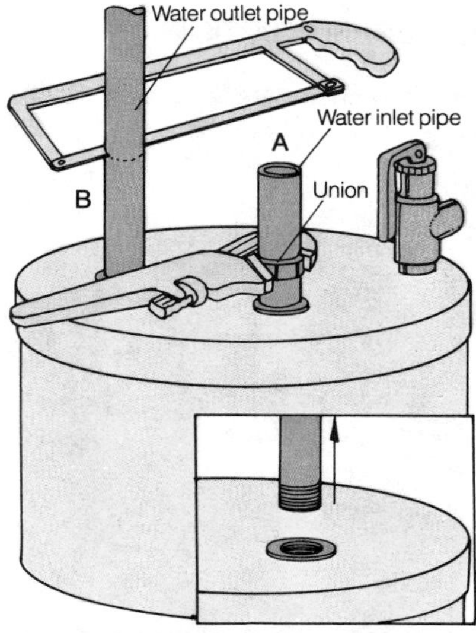

Drawing 60. Disconnect the water supply pipes by unfastening a union (A) or cutting through the pipe (B).

Electric water heaters

Capacity in gallons: 40, 50, 60, 70, 80, 90, 100

Number of bathrooms: 1 (2 people), 1½ (3 people), 2 (4 people), 3 (5 people)

Gas water heaters

Capacity in gallons: 40, 50, 60, 70, 80, 90, 100

Number of bathrooms: 1 (2 people), 1½ (3 people), 2 (4 people), 3 (5 people)

Gallon capacities on the graphs are based on households with two water-using appliances; a household without a dishwasher or a washing machine will need a slightly smaller tank. Water heater needs will also vary if there are more or fewer people in the household than shown in parentheses. Some people need more water: with small children, for example, more hot water is required for laundry.

disconnect the fuel supply pipe from the inlet valve. You'll also need to remove the draft diverter from the flue pipe of a gas heater (see **Drawing 61**).

Most water heaters have temperature and pressure relief valves to prevent explosions in case the heating mechanism fails. The valves are inexpensive, and it's a good idea to get a new one when you get a new heater. You may be able to use the existing overflow pipe (see **Drawing 62**).

Installing a new heater. Remove the old heater and set the new one in place. Check that the heater is plumb and level (see **Drawing 63**); shim if necessary.

Plumbing the tank. If the new tank is a different height than the old one, you may have to tinker with the plumbing to make everything come together properly. Use flexible pipe connectors (see **Drawing 64**) or unions to hook up both the water and gas lines. The connectors (lengths of flexible tubing) simply thread onto the pipe and bend as needed to make the hookup. If the pipes aren't threaded, replace them with threaded nipples and secure the connectors to them with a wrench.

Activating the heater. For an electric heater, run metal-clad electrical cable from the power source. With all the connections made, open the water inlet valve to the heater. When the tank is filled with water, "bleed" the supply pipes (open the hot water faucets to allow air to flow out of the pipes).

Test the temperature and pressure relief valve by squeezing its lever. Open the gas inlet valve or energize the electrical circuit to fuel the heater. For gas heaters, light the pilot according to instructions (usually on the control panel plate). Adjust the temperature setting as desired.

Finally, check all connections for leaks. If you're working on a gas heater, brush soapy water on the connections (see **Drawing 65**)—bubbles indicate a gas leak.

For solutions to water heater problems, see pages 38–39.

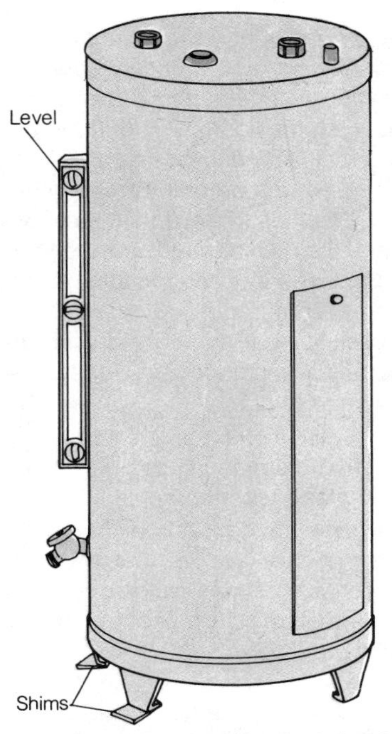

Drawing 63. Use a level to check that the new heater is plumb; if necessary, shim with thin scraps of wood.

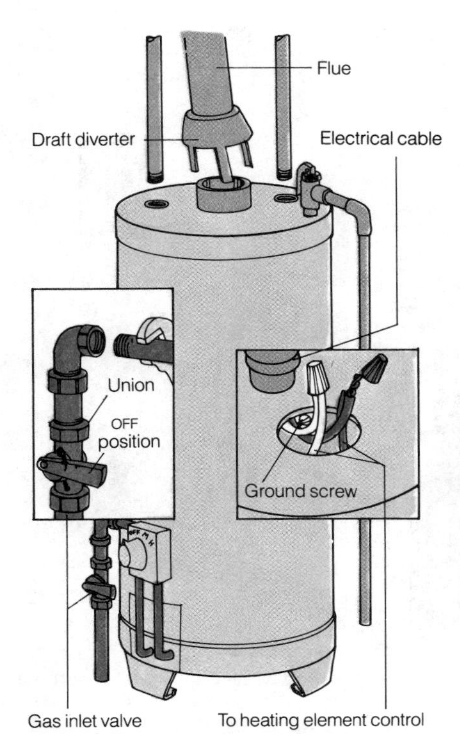

Drawing 61. Detach the fuel supply pipes (or electrical cable); then, on a gas heater, remove the draft diverter.

Drawing 62. Unscrew the overflow pipe from the relief valve on the old unit to use on the new water heater.

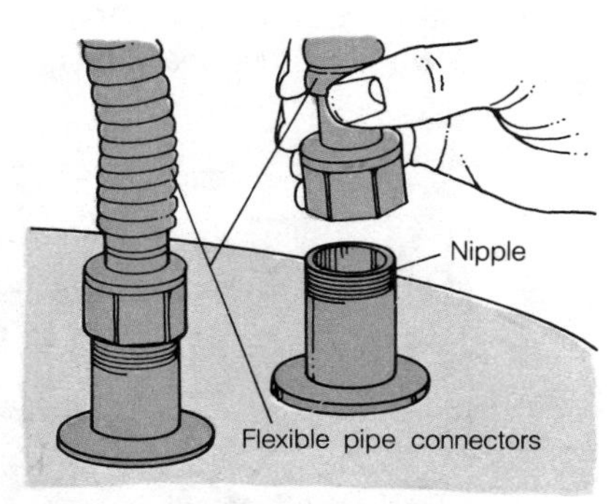

Drawing 64. Use flexible pipe connectors to simplify water and gas line hookups to the water heater.

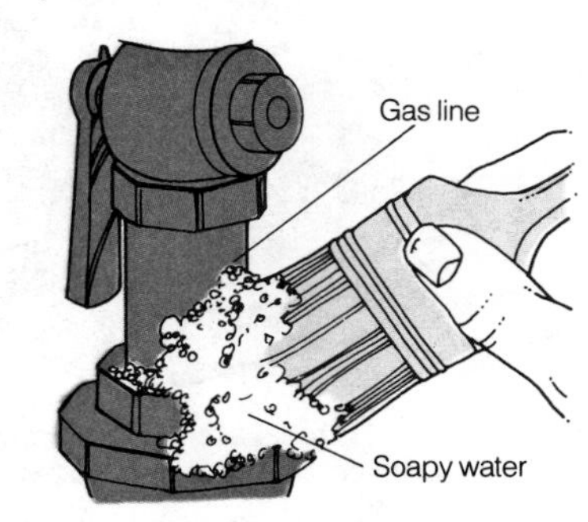

Drawing 65. Check all gas line connections for leaks, applying soapy water with a brush; bubbles indicate a leak.

Heating water with the sun

Solar energy—the sun's natural warmth—is an effective source for heating water. Before 1985, solar heating systems flourished, due largely to federal tax credits for their installation. But when the credits were eliminated in 1985, such systems lost their luster, since they were too expensive to realize savings in a reasonable period of time. Today, spiraling energy costs, technological advances, and lower prices are making them a viable alternative once again.

Solar water heaters fall into two general categories: passive and active. Along with the sun's energy, active systems depend on thermostats, fans, pumps, and valves powered by electricity; passive systems don't need any mechanical components or conventional energy.

Passive systems

There are two basic passive systems for domestic water heating: batch water heaters and thermosiphoning water heaters. Both are simple, low-cost ways to introduce solar energy into your home.

Batch water heaters. Often called "breadbox" water heaters because of the shape of their characteristic containers, batch water heaters require no energy input or specialized hardware to make them work; they just sit in the sun and get hot.

In a batch heater, the collector and storage components are one and the same—often just a glass-lined water heater tank stripped of its outer casing, insulation, and heating mechanism. The tank is painted flat black to absorb solar radiation, and is housed in an insulated box that's glazed on one side and oriented within 30 degrees of true south in an unshaded location—usually at ground level (see **Drawing A**). The glazing must also be tilted up at an angle from the horizontal that matches your latitude. Because a full tank is very heavy, roof mounting is inadvisable unless a structural engineer determines that it's feasible.

Most systems employ one or two 30 or 40-gallon tanks to preheat water on its way to a conventional heater. But in some regions, systems with two or more tanks connected in series can displace a conventional heater altogether during the sunniest months.

Freezing may be a problem in cold climates; in such areas, it's often necessary to drain the system in winter.

Recent developments in batch heaters include the use of special glazings, reflectors, and tank coatings to increase solar efficiency. The performance of such "enhanced" batch heaters can approach that of more sophisticated systems, and their cost effectiveness is unrivaled.

As with all solar water heaters, batch heaters may require some changes in domestic habits if you are to reap their full benefits. Since peak water temperatures are usually achieved in midafternoon, it's best to schedule your maximum use of hot water for this time. However, special tank coatings and insulated covers can help maintain high water temperatures.

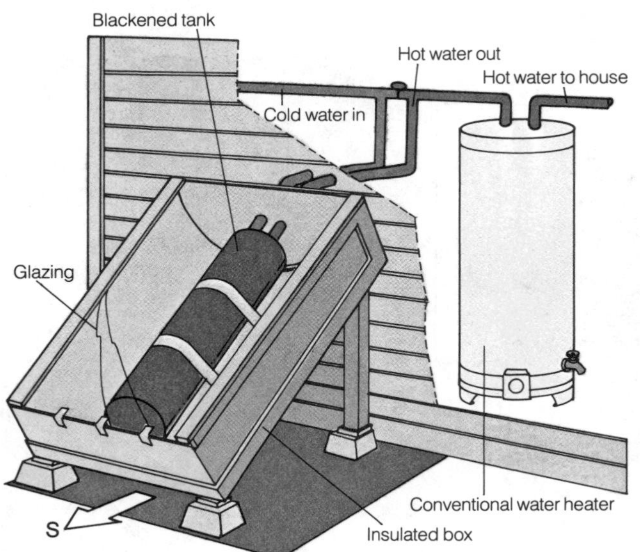

Drawing A. Batch water heater preheats water for a conventional heater; its darkened tank serves as both collector and storage.

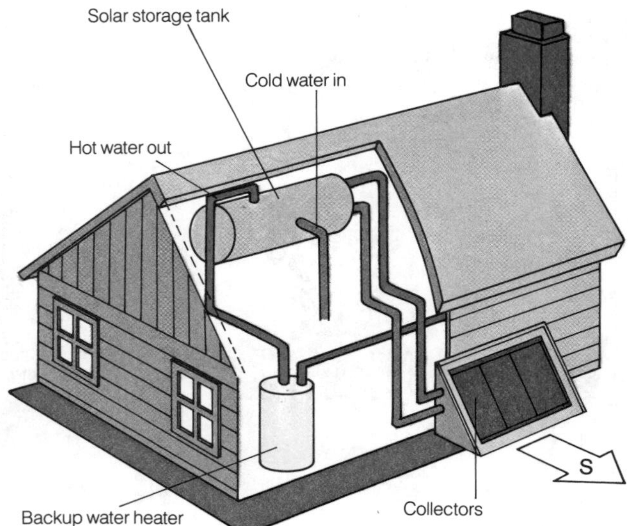

Drawing B. In a thermosiphoning water heater, water warmed in collectors rises by natural convection to tank set inside house; hot water is drawn off through conventional heater.

For construction help and additional information, consult a solar heating specialist.

Thermosiphoning water heaters. In these passive heaters, water circulates from solar collectors to a solar storage tank by natural convection; the water rises in response to the sun's heat just as air does. To initiate this convective current, the collectors are mounted with their tops below the bottom of a well-insulated tank into which they feed (see **Drawing B**). The solar-heated water is drawn off for household use indirectly—through the conventional heater, which acts as a backup to the solar water heater.

The system just described is called an open-loop system, because plain household water runs directly from the collectors into the storage tank. To prevent wintertime freezing, the collectors must be drained when the temperature drops.

To eliminate the problems of water freezing in a solar heating system, you can use a closed-loop system that employs a heat exchanger to heat water. In this system, a mixture of water and antifreeze gathers heat from solar collectors and circulates through a closed loop of tubing to a heat exchanger immersed in a special hot water storage tank.

The heat exchanger, usually coiled copper tubing or a finned device similar to a car radiator, is double-walled to prevent leaking. It keeps the antifreeze mixture separate from your water (see **Drawing C**). The coiled or finned design enhances heat exchange by providing a large area of contact between the two fluids while keeping them safely separated.

Active systems

With an active solar water heater, the collectors are usually mounted on the roof, and the storage tank is placed at floor or basement level. With this arrangement, you'll need pumps, valves, and automatic controls to circulate regular household water (in an open-loop system) or antifreeze (in a closed-loop system) to the collectors and back again (see **Drawing D**).

Though some active systems are custom-built, most are available as kits for professional installation. These packaged systems include collectors, tanks, thermostats, pumps, and piping. If you have plumbing and wiring skills, you may be able to install one of these yourself on a house with a south-facing roof. Even a shed or garage roof will do: often only 1 to 2 square feet of collector area are needed per gallon of water to be heated.

These kits are less expensive than custom-built models, but they're not exactly simple to install. You'll need quite a bit of expertise and derring-do—and a good dose of patience. You'll probably have to pass a building inspection, too. Be sure before buying a kit that the instructions are adequate, or that the seller will help you if they aren't. Unless you're an experienced do-it-yourselfer, it's best to have your system installed by an expert, even if you have to pay more.

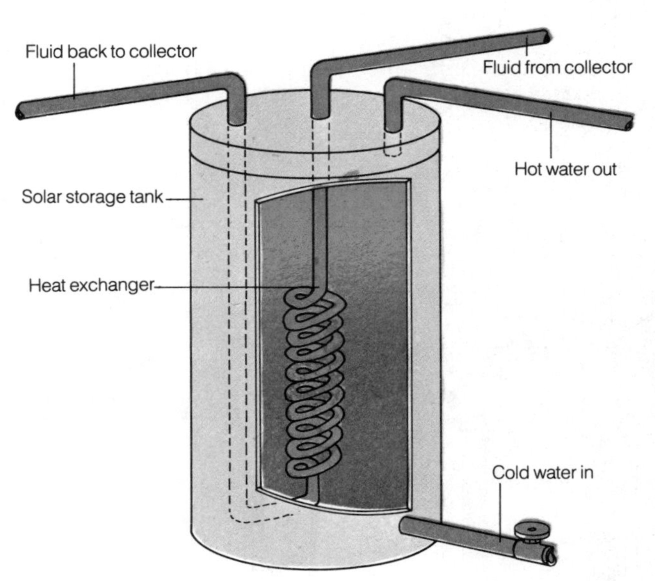

Drawing C. Heat exchanger immersed in storage tank provides a large surface area for transfer of heat, while keeping heat-transfer fluid separate from potable water.

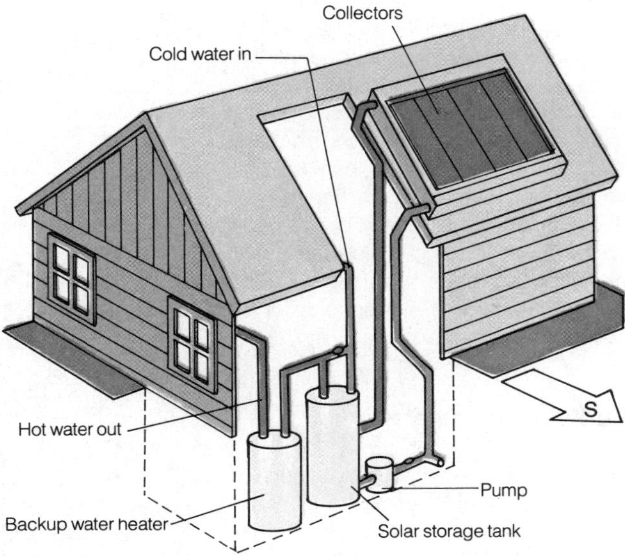

Drawing D. In an active solar water heater, pump circulates water (or antifreeze) from collectors to storage tank and back.

Tying in Water Softeners & Filters

Chemicals in water—a nemesis of homeowners—can be both a nuisance and a hazard, especially when they leave mineral deposits in supply pipes, give water an unpleasant taste and smell, leave a hard-water ring around the tub, and cause health problems. Tap water that contains rust, chlorine, sulfur, or other unwanted substances may need to be treated with a water softener or purified with a water filter.

Softening your water

All water softeners operate by substituting sodium (salt) for calcium, magnesium, or iron, any of which can cause hard water. A water softener not only eliminates soap scum but also prevents the buildup of harmful minerals in such water-using appliances as water heaters.

But adding sodium to the water supply can be a potential health problem for people who need to restrict their intake of salt. One way to avoid this problem is to hook up the softener only to the hot water pipes—those used for bathing and washing.

The softening unit attaches to the main supply pipe just past the point where the water enters the house. To install the unit, you'll need to tap into the main supply pipe in one of two places. You can either install the softener before the hot and cold water pipes branch off, so that all the water is softened, or you can install it only on the hot water pipe, a health-wise decision. In either case, check local codes for extending piping (see "The planning sequence," page 60). Follow the manufacturer's instructions for installing the softener.

Purifying your water

The various types of water purifiers on the market today are each effective against different kinds of contaminants. The three main types are distillers, activated carbon filters, and reverse-osmosis purifiers. A fourth type of water treatment—ultraviolet—is sometimes incorporated into a system to kill microorganisms.

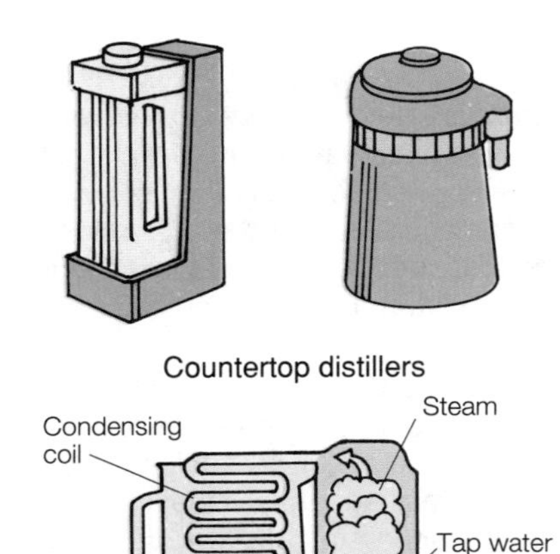

Drawing 66. Countertop distillers boil water, turning it to steam. Cooling vapor condenses and is stored as clean water in a separate chamber.

Distillers. Distillers kill bacteria and remove such minerals as magnesium and calcium. But they do not remove contaminants that evaporate with water, such as solvents, pesticides, and gases.

Most distillers are containers that sit on the countertop (see **Drawing 66**). You fill the distiller with water, which is then boiled, causing steam to rise. The steam condenses and collects in a separate chamber. You pour or pump out the purified water when you need it.

Most distillers are portable and don't require a plumbing hookup. However, they have some important disadvantages: For each gallon of clean water they produce, distillers can consume up to 3½ kilowatt hours of electricity and take up to 5 hours to complete the job.

Activated carbon filters. As water passes through an activated carbon filter, solid particles, some organic contaminants, and inorganic chemicals are captured and held. But carbon filters don't remove dissolved solids or bacteria. In fact, they can incubate bacteria, which means that filter cartridges must be replaced at least twice a year, depending on use.

If a filter sits unused for more than a few days—such as while you're on vacation—always let the water run to purge bacteria.

Carbon filters come in several styles: beneath-the-counter, countertop, and faucet mounted. Because a carbon filter's effectiveness is directly related to the amount of activated charcoal it contains, small, faucet-end models are relatively ineffective.

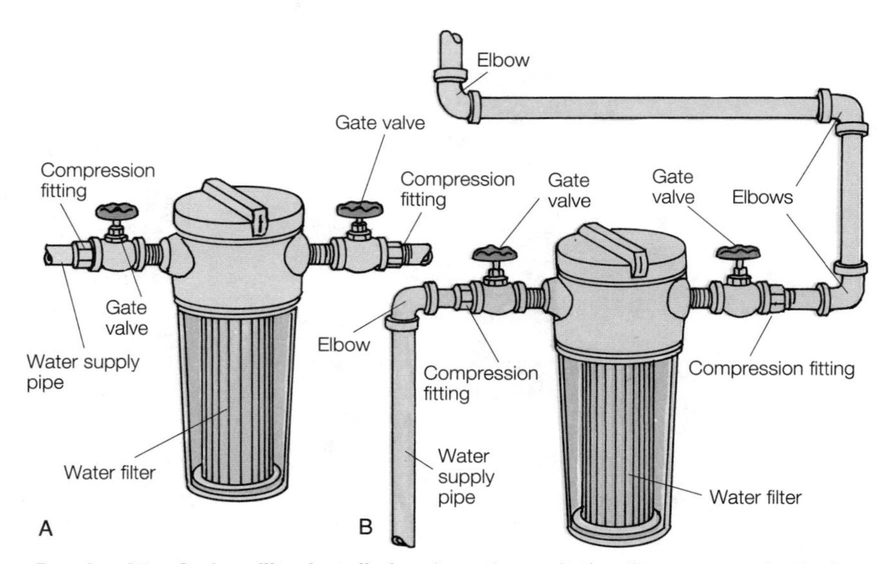

Drawing 67. Carbon filter installation depends on whether the water supply pipe is horizontal (A) or vertical (B). Installing two gate valves allows you to shut off the water supply and control backflow when you change the filter core.

Carbon filters are tied directly into the plumbing, usually beneath the kitchen sink, and don't require any electricity. When it's time to change the filter core, you simply close the gate valves to shut off the water, unscrew the filter body from the cap, and replace the used filter with a new one.

Because the filter must be installed in an upright position, the installation method depends on whether you're tying into a horizontal or vertical line.

- *Installing in a horizontal pipe.* To attach a water filter to a horizontal pipe (see **Drawing 67A**), cut out a length of pipe where the filter will be installed. Thread a gate valve onto a nipple at each side of the filter and attach the valve with compression fittings (or unions if the pipe is galvanized).
- *Installing in a vertical pipe.* If the filter is being installed in a vertical pipe (see **Drawing 67B**), cut out a section of pipe, install piping with four elbows as shown, and attach the filter in the lower horizontal leg. Attach gate valves as described above.

Reverse-osmosis purifiers. Reverse-osmosis systems remove all sorts of impurities from water, which is why they're used extensively in hospitals and laboratories. They work by forcing pressurized water through a membrane that blocks everything but chemically pure water molecules (see **Drawing 68**).

Often, a purifier is used in combination with carbon filters that remove sediment before it reaches the membrane and enhance the taste of the water after it leaves.

The clean water is stored in a tank, normally located under the sink, and is delivered through a separate, small faucet. The water may also be routed to a hot water dispenser and an icemaker. Polluted water is delivered to the sink drainpipe.

- *Hooking up a reverse-osmosis purifier.* The purifier's faucet typically fits in a hole at the rear of the sink rim. If your sink isn't equipped with a hole or knockout for this purpose, you may be able to have a plumber drill a hole for you. An easier and less expensive solution is to mount the faucet in the countertop, letting the faucet overhang the sink.

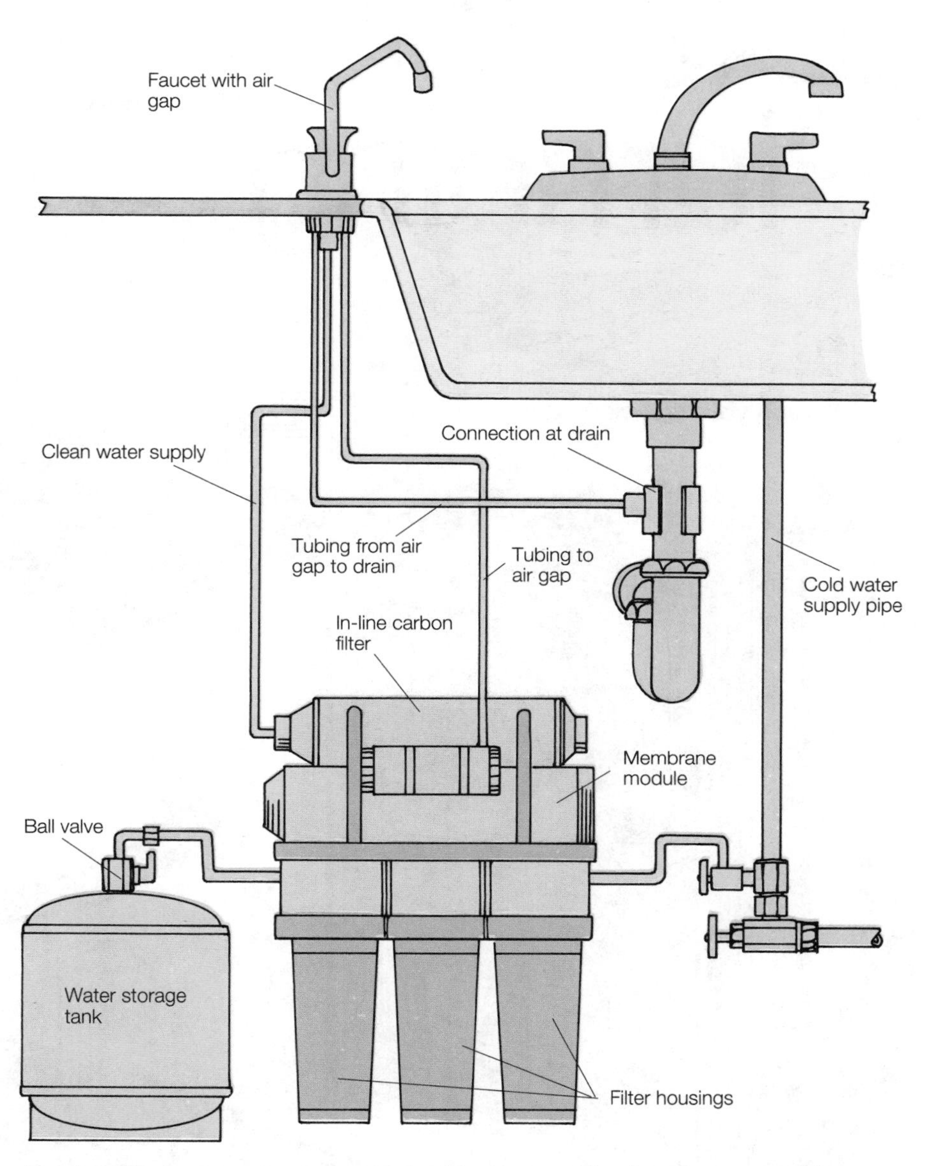

Drawing 68. Reverse-osmosis water purifier connects directly to the water supply pipe. Water travels through a prefilter, a reverse-osmosis membrane, and a postfilter. Clean water is held in an undersink storage tank and is delivered through a separate faucet.

The filter unit is usually located under the sink with the tank. Connections are made with flexible plastic tubing (see pages 47–48) or with soft copper supply pipe (see page 52). Follow the manufacturer's instructions for installation and for connection to the cold water supply pipe (also see Drawing 53 on page 82).

- *Hooking up to a hot water dispenser and icemaker.* One purifier can route clean water not only to the sink but also to a hot water dispenser and icemaker.

All you need to do is add a compression tee to the tube that carries clean water from the filter and route flexible plastic tubing or soft copper supply pipe to each additional appliance. For more information about connections, see "Adding a Hot Water Dispenser," page 82.

Exterior Plumbing

Adding an Outdoor Faucet

Adding a faucet, or hose bibb, on the exterior wall of your house is a fairly straightforward procedure that will pay off in convenience when you need a handy source of water outside your house.

Several faucet models are available. Nearly all have a threaded spout for attaching a hose. Some have bodies with female threads and are screwed onto a pipe; others have male threads and are screwed into a threaded tee or elbow. One type fits onto horizontal pipes, another onto vertical pipes.

Some hose bibbs are made with a notched flange, or escutcheon, that allows the faucet to be screwed to an exterior wall (this type of outdoor faucet is also referred to as a sillcock).

Placing the faucet. First decide where you want to install the new faucet. It should be convenient for outdoor watering and, if possible, high enough on the wall to clear a bucket.

Be sure to consider, too, the location of the indoor cold water pipe you'll be tapping into (it's probably in your basement or crawlspace). As with any plumbing project, you'll need to plan carefully how you'll tap into the pipe and have ready all the pipe and fittings necessary for the job before you begin (see "Pipefitting Know-how," pages 46–65).

Drilling into the wall. Before you start drilling, check indoors to make sure you won't hit any obstructions, such as drainpipes, electrical conduit, heating ductwork, studs, or floor joists.

Be sure to avoid the foundation. If the water supply pipe is located below the foundation's top surface, plan to drill above the foundation and route the new pipes down to the water supply pipe. If possible, drill a small pilot hole from the inside out to mark the right location.

Select the correct bit for the job: a spade bit for wood, a masonry bit for brick, concrete, or stucco. Then, using an extender if necessary, drill all the way through the wall from the outside, boring a hole just large enough to accommodate the pipe that will be attached to the faucet.

Connecting the faucet. Turn off the water at the main shutoff valve (see page 23) and drain the pipes.

The quickest way to tap into a pipe is with a saddle tee, but for a more professional job, use threaded or soldered fittings where appropriate (see **Drawing 1**). You may want to add an indoor shutoff valve (see pages 70–71), an extra convenience anywhere but a must in cold-winter areas, unless you're using a freezeproof faucet.

Finally, run your new pipe through the wall and connect the faucet to it. Be sure your connecting pipes are well anchored to the house's framing (studs and joists) near the wall, as well as all along the pipe run.

Finishing the job. When you have everything connected, fill any gaps around the pipe, both inside and out, with waterproof silicone caulking or foam sealant. When installing a flanged faucet, you may want to caulk the space around the pipe before screwing the flange in place (the caulking compound will spread behind the flange to form a seal).

Some codes require that you install a vacuum breaker, also called a backflow preventer (see **Drawing 2**). This device, which you simply screw on between the faucet spout and the hose itself, prevents the backflow of polluted water into your home's water system. Check with your local building department for the requirements in your area.

Installing a freezeproof faucet. If you live in an area where winter temperatures often dip below freezing, it makes sense to install a freezeproof faucet (see **Drawing 3**). This type of faucet has an elongated body that extends well into a basement or crawlspace and a valve seat located far back into the body. When you turn off the faucet, the water flow stops back inside the house.

Freezeproof faucets are self-draining. You install the unit at a slight tilt toward the ground outside, which allows any water remaining in the body after the faucet is turned off to run out of the spout.

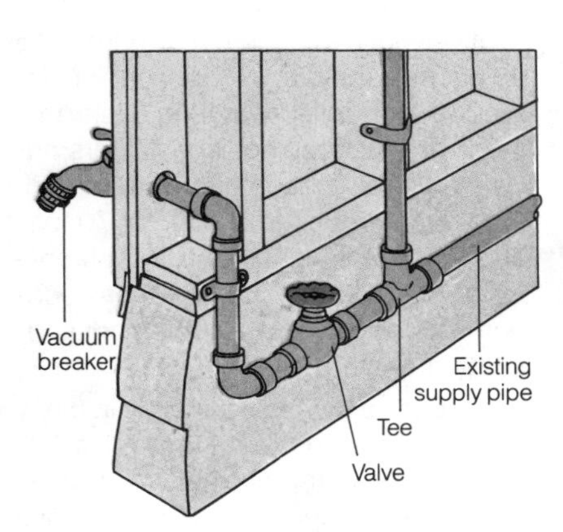

Drawing 1. Tap into the cold water supply pipe to connect an outdoor faucet.

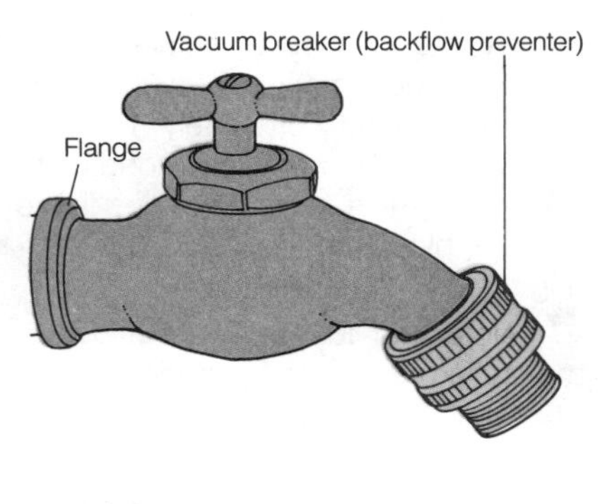

Drawing 2. A vacuum breaker prevents the backflow of used water into the house's water system.

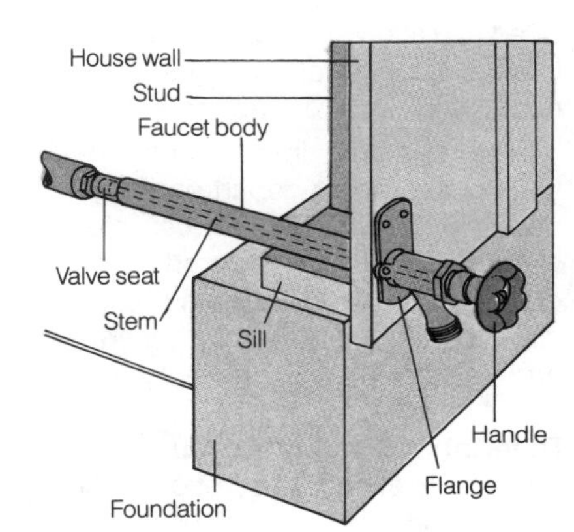

Drawing 3. A freezeproof faucet has a long body that extends into the basement or crawlspace.

Planning a Garden Irrigation System

Designing a successful irrigation system, whether it's a sprinkler system, drip irrigation, or a combination of the two, requires knowledge of hydraulics, soil types, plant growth needs, and irrigation equipment. But don't be discouraged—planning and even installing a system is well within the ability of most homeowners. Instructions for installing a sprinkler system begin on page 94; for information on drip systems, see page 98. To convert an existing sprinkler system to a drip system, turn to page 101.

For help with particular problems or for expert advice, look in the Yellow Pages under "Irrigation Systems & Equipment," "Landscape Contractors," or "Sprinklers—Garden & Lawn."

Mapping your property

To design a watering system, you'll need to prepare a fairly detailed scale drawing of your property, determine the water-retaining characteristics of the soil, and measure water pressure and water flow rate.

Making a scale drawing. Use graph paper to make a scale drawing of the areas to be watered, marking locations and types of plants, any special water needs they have, and whether they're deep or shallow rooted. Pencil in water sources and structures and such obstacles as fences, walkways, and patios. Also note any slopes and elevation changes, since they can affect water distribution. The more detailed and accurate your plan, the easier it will be to select the right components when you're ready to shop.

Now is the time to check with your local building department for any necessary permits. Also, investigate the various products on the market; if you decide on a particular brand, pick up copies of the manufacturer's literature and workbook.

Determining soil type. Your garden's soil greatly affects absorption, evaporation, and the lateral movement of water. If you're not sure of your soil type, wet some and squeeze it into a ball. If it crumbles, your soil is sandy. If some of the ball holds its shape, you have loam. And if it sticks firmly together, it's clay.

Sandy soil readily accepts water but doesn't retain it well. Loam accepts and retains water. Clay doesn't readily accept water, but once it does, it retains it well.

Sprinkler systems work well in sandy soil and in loam. Clay soil holds so much moisture that excessive runoff and fungus occur when conventional sprinkler heads are used, so choose low-gallonage heads.

Measuring water pressure. Most sprinklers won't operate efficiently if the water pressure, measured in pounds per square inch (psi), is too low. To measure your home's water pressure, borrow or rent a water pressure gauge (see **Drawing 4**) from a hardware store, a plumber, or a tool rental company. Screw the gauge onto an outdoor faucet and, with all other water outlets turned off, turn the faucet on full. Record psi at each outside faucet location, taking several readings throughout the day, and use the lowest pressure.

As a rule, sprinklers work best with high water pressure. Drip systems need lower pressure, so if pressure is over 75 psi, you'll need to install a pressure regulator (see Drawing 14 on page 98).

Measuring water flow rate. Flow rate, the amount of water that moves through pipes in a given period of time, is measured as gallons per hour (gph) or gallons per minute (gpm).

To determine flow rate, place a 1-gallon container under an outdoor faucet and count how many seconds it takes to fill the bucket completely. Then divide the total number of seconds into 60 to determine gpm. Write this figure on your plan; you'll use it when plotting circuits.

Note that to use this method of measuring gpm, the outdoor faucet must be the same diameter as your service line.

Both sprinklers and drip irrigation emitters have designated flow rates. Generally, the total output of a circuit of sprinklers or emitters should not exceed 75 percent of your plumbing's available water flow at the faucet; otherwise, the heads or emitters won't operate properly, and household water pressure may dip. The solution is to create several separate circuits, each directed by its own control valve.

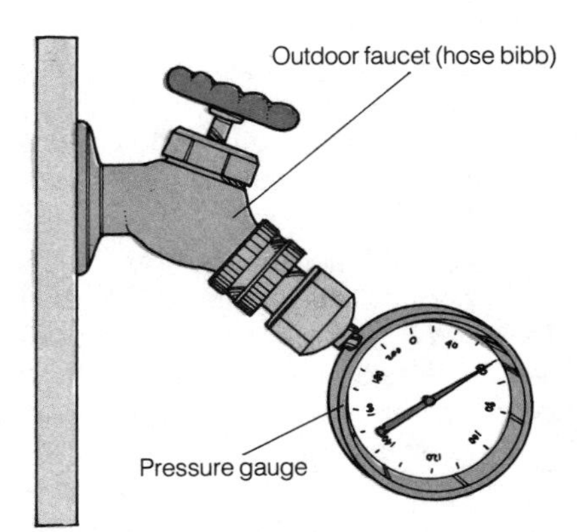

Drawing 4. To determine water pressure, screw a water pressure gauge onto an outdoor faucet and turn water on full.

Positioning sprinklers & emitters

Planning your system on paper, whether it's a sprinkler system or drip irrigation, will help you think the system through, guide you in ordering materials, and serve as a permanent record of where any pipes or tubing is buried.

Locating sprinkler heads, circuits, and pipe runs. The two broad categories of sprinkler heads are spray and rotary. Spray heads operate at relatively low water pressure, from 15 to 30 psi; they're a good choice for precise, controlled watering of shrubs, irregular landscaping, and fairly small lawns. Rotary heads need more pressure to operate (from 30 to 70 psi) and throw water substantially further—up to 90 feet; this makes them economical for very large lawns and landscaped areas.

Sprinkler heads also come in many different spray patterns, including full, half, and quarter circles, as well as rectangular shapes. Some heads have adjustable patterns and throw distance. Low-precipitation-rate nozzles reduce runoff, improve spray uniformity, and allow a larger area to be irrigated with a given amount of water.

For open lawn areas where foot traffic and mowing will occur, install pop-up heads that automatically rise when the water goes on and drop down when watering is finished.

On a copy of the scale drawing of your property, detail your proposed sprinkler system (see **Drawing 5**). Begin by noting where you need to locate sprinkler heads. To establish the spacing and correct water distribution pattern for each head, check the manufacturer's workbook to find out the radius, or throw, of each type of head. Use a compass to draw the rounded spray coverage patterns, making sure they overlap sufficiently to provide complete coverage.

Next, break your system into separate circuits, or stations, keeping in mind the flow-rate limitations discussed on the facing page. Group the heads by sprinkler type and water requirements: don't place rotary and spray sprinklers, shrub and lawn sprinklers, or low-gallonage and standard sprinklers on the same circuit. Remember that when a circuit goes on, all the sprinklers along its line will deliver water simultaneously.

Starting at the control valve for each circuit, sketch in the piping that will connect the sprinklers on the same circuit. Try to avoid running pipes under paved areas. Note that a T or H-shaped circuit will deliver water more evenly to all heads than a straight-line circuit (the last head in a line typically receives less water pressure).

Choosing and placing emitters. Decide on the correct gallonage and number of emitters for each plant, depending on your soil type (see the chart on page 100). In general, use higher-gph emitters for trees and plants in sandy soil, lower ones for shallow-rooted plants and in clay soil. Space emitters closer together for shallow-rooted plants (see **Drawing 5**).

Group plants on separate circuits according to their water needs and root depths. If possible, place trees, shrubs, flowers, vegetables, and containers all on different lines so you can schedule watering to suit their individual needs.

For most gardens, plan on using ½-inch polyethylene tubing for lateral lines to shrubs and trees. Be careful not to run the line too long or put too many emitters on it, especially if your system is extensive: the tubing has limits on how much water it can handle efficiently. Be aware that running a line uphill shortens the possible run; running it downhill increases it.

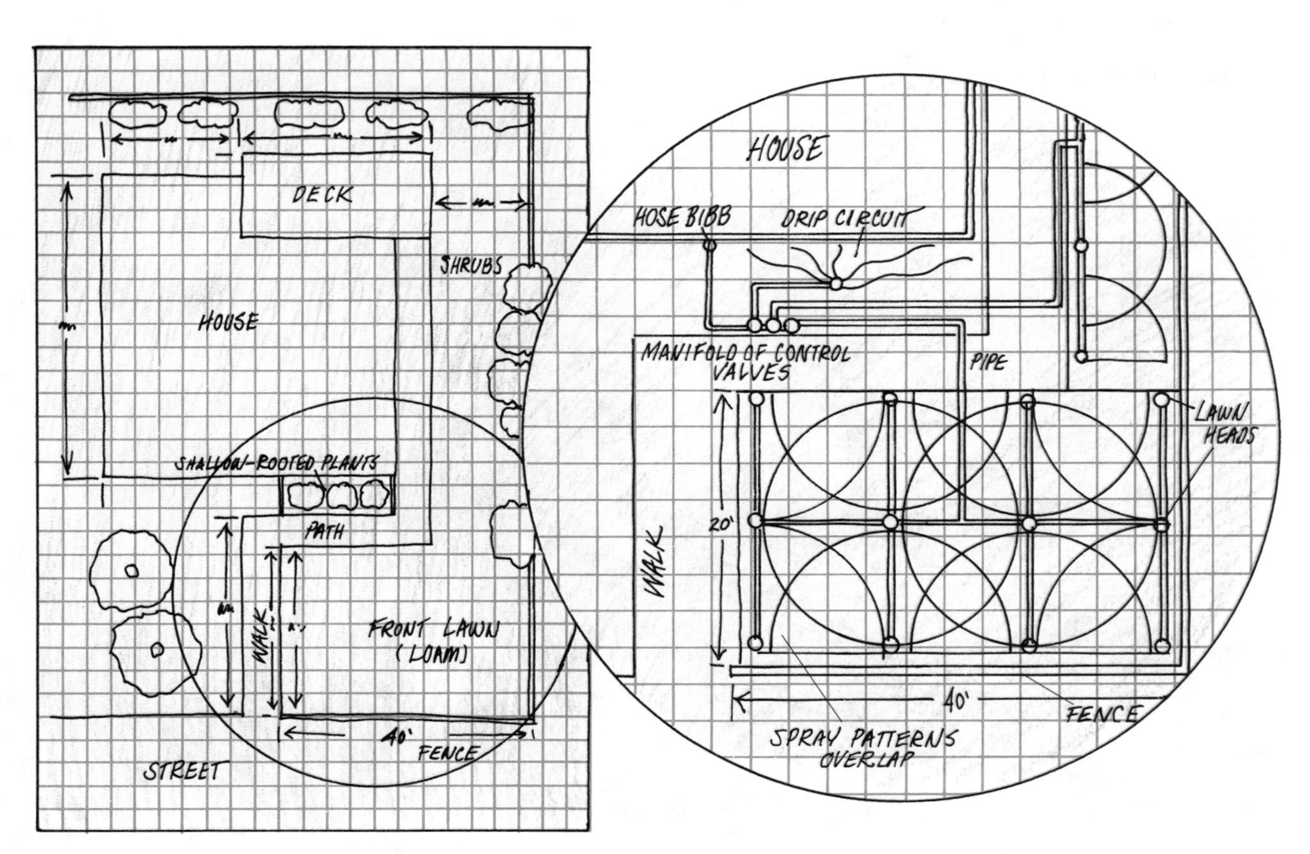

Scale drawing of property

Detail of proposed circuits

Drawing 5. Planning an irrigation system

Installing an Underground Sprinkler System

The most efficient way to water a lawn is with an underground sprinkler system. If you're putting in a new lawn or reseeding an old one, this is a good time to install such a system. You can also add an in-ground system to an existing landscape, but you'll have a lot of repair work to do on your lawn afterwards. Keep in mind that for watering plants, drip irrigation is more efficient; for information on drip systems, turn to page 98.

System components

The components you'll need to install a system include antisiphon control valves, essential for preventing tainted water from flowing back into your home's water supply; pipes and pipe fittings, usually made of PVC (polyvinyl chloride); risers; sprinkler heads; and a controller, or timer. (Typical components are illustrated in **Drawing 6.**)

The system begins at a cold water supply pipe, where you connect the new supply line with a tee or compression tee fitting. Locate a shutoff valve on the new supply pipe so you can shut off the sprinkler system without turning off the water to your house.

The new supply pipe carries water to control valves with integral antisiphon devices. Each valve operates a circuit, or a separate set of sprinklers. Place the valves in a convenient, inconspicuous place, grouping them into what's called a manifold to make operation easier and avoid extra digging.

In an automated system, low-voltage wires run from the manifold to a controller, which can be anything from a simple mechanical timer to a complex digital system. The controller directs the watering cycle by automatically activating the control valves for the different circuits so they turn on when and for how long you wish.

Installation

Plumbing with PVC pipes and fittings is not difficult; for help working with plastic pipe, see pages 47–51. Make sure that your scale drawing shows the location of the manifold and all pipes and sprinkler heads.

Trenching. The first task is to dig V-shaped, 8-inch-deep trenches for burying the pipe. To locate the trenches, lay the pipes on the ground according to your plan and, using a shovel, mark their location on the ground. You can dig the trenches with a flat spade or with a rented trenching machine. To salvage any sod, gently work the spade beneath the sod layer and peel the sod away before digging deeper.

Connecting to the supply pipe. Begin pipe installation where you will tie into the cold water supply pipe. If that pipe is 1-inch diameter or larger, you'll need to run 1-inch pipe to the valves; if it's only ¾-inch diameter, run ¾-inch pipe.

CAUTION: Before doing any work on the supply pipe, shut off the water at the main shutoff valve (see page 23). Open a faucet at the low end to drain the pipes.

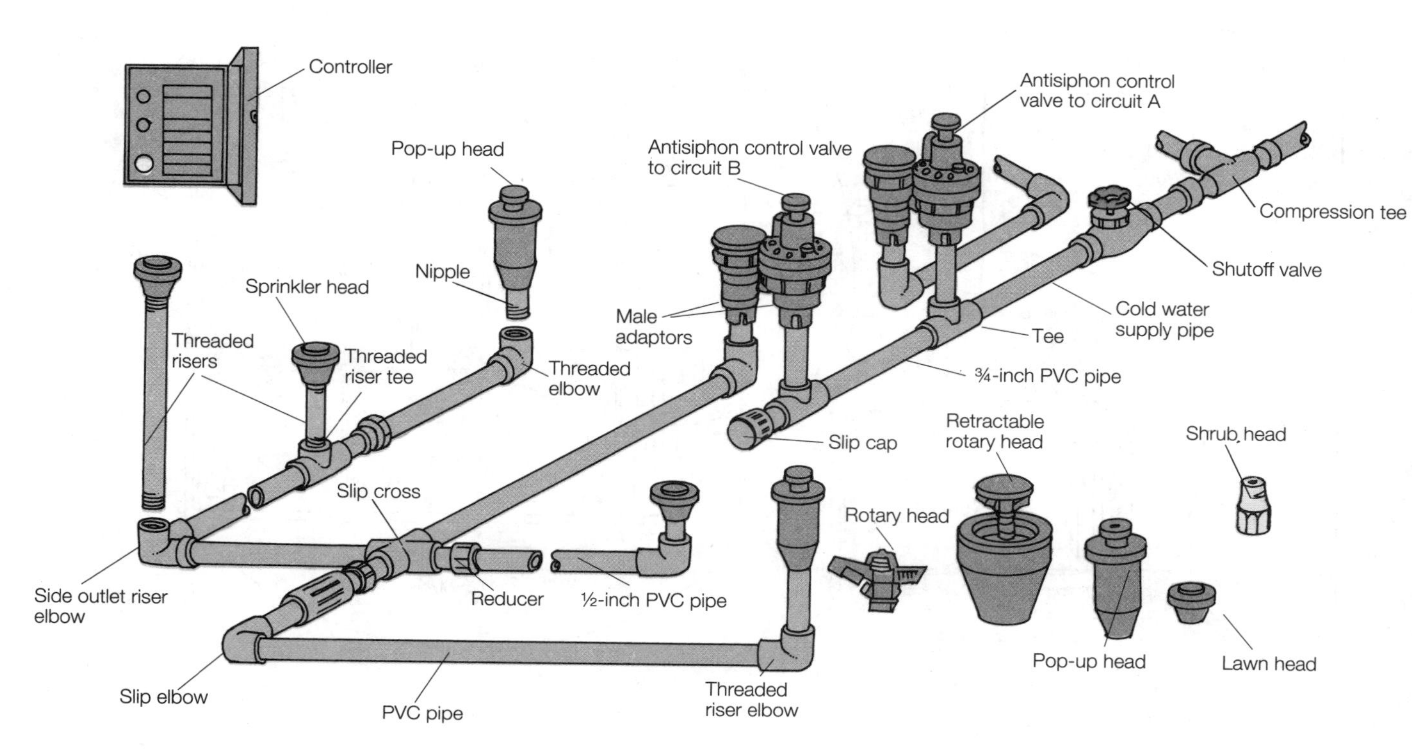

Drawing 6. Typical sprinkler system components

In cold-winter areas, you'll need to tap into the cold water supply pipe inside the house (see **Drawing 7A**). Tie into the pipe with a tee fitting and install a shutoff valve beyond the tee (choose a stop-and-waste valve so you can drain the system at this point). To run the new line to the manifold, drill a hole through the siding above the foundation.

In milder climates, you can tap into an outdoor faucet (see **Drawing 7B**). Simply remove the existing faucet, add a tee, and then attach a new faucet to the tee. Install a shutoff valve, as shown.

If there is no nearby faucet, tap into the main supply pipe (see **Drawing 7C**) before it enters the house (using a compression tee makes this job solder-free).

Once you've installed the shutoff valve, you can turn the gate valve at the water meter back on (leave the new shutoff valve turned off).

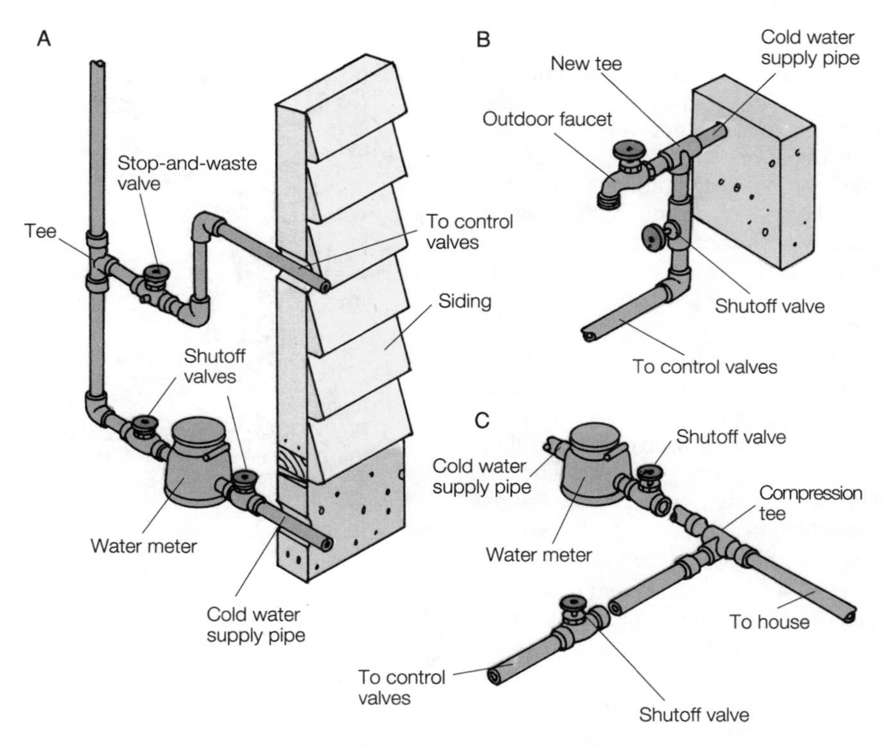

Drawing 7. Tap into the cold water supply pipe inside the house in cold-winter areas (A). In milder climates, tap in at an outdoor faucet (B) or along the length of the main supply pipe (C).

Adding the control valves. Run pipe from the shutoff valve to the manifold; flush out any dirt. To install each control valve, you'll generally need two ¾-inch male-threaded adapters; wrap the threads with fluorocarbon (pipe-wrap) tape and screw the adapters into the valve; hand-tighten. Cement the plastic-pipe riser into the tee or elbow in the line and into the adapter (see **Drawing 8**). Most codes require that control valves be at least 6 inches above the ground and above the highest head in the circuit.

Laying out piping. From the control valves, run the pipes for each circuit (see **Drawing 9**). Before cementing each threaded riser tee or elbow in place, screw in a riser so you can align the fitting properly with the surface of the ground—risers should always be at a 90° angle to the ground. Don't add the sprinkler heads yet.

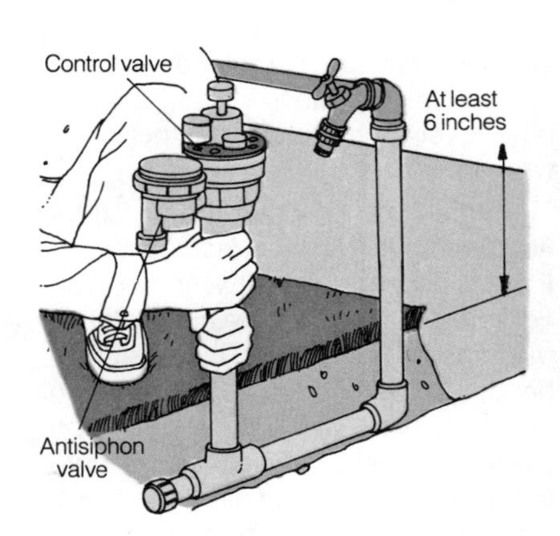

Drawing 8. Attach an antisiphon control valve to the water pipe, making sure the valve is at least 6 inches above the ground.

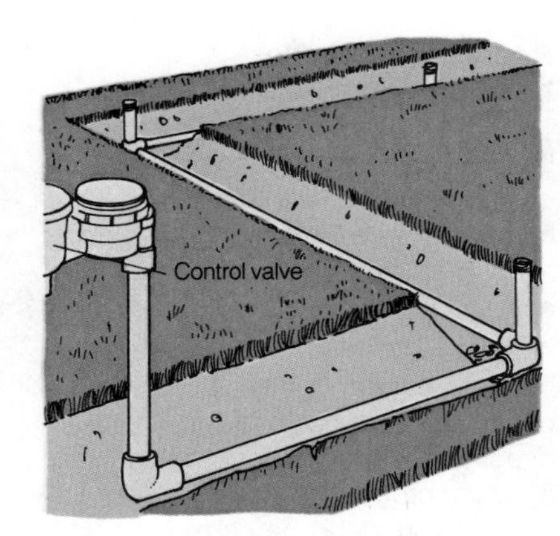

Drawing 9. Connect the system from the control valve outward. Before attaching the heads and filling the trenches, turn the water on to check for leaks and then fill in the trenches.

Finishing up. When all piping is assembled, allow the solvent-cement to cure for at least an hour. Open the shutoff valve that supplies the sprinklers. Next, briefly open the control valves to purge any dirt from the pipes. Wrap riser threads with fluorocarbon (pipe-wrap) tape, screw on the sprinkler heads, and adjust their positions. Turn the water on again to check for proper operation and any leaks.

To hook up the controller, run low-voltage cable (typically direct-burial AWG-14) from the valves through the trenches to the controller's location (normally a garage or other place where you can plug it into a 120-volt circuit). To program the controller, follow the manufacturer's instructions.

When the system is working properly, backfill the trenches.

Fixing Faulty Sprinklers

Normal operation can cause sprinklers to clog, jam, leak, or spray improperly. And, of course, they can be easily damaged by lawn mowers, car tires, foot traffic, and the like. Faulty sprinklers can not only damage plants but also waste huge amounts of water. That's why it's important to inspect them at least once a month during the watering season, more often if they're exposed to hazards.

Fixing or replacing broken or damaged parts is usually easy: few special tools are required, and replacement parts are readily available in stores that specialize in irrigation equipment. To find dealers, look in the Yellow Pages under "Irrigation Systems & Equipment" and "Sprinklers—Garden & Lawn."

Because many different manufacturers make sprinkler equipment, take the defective part with you to the dealer so you can get a correct replacement. Difficult repairs, such as problems with automatic controllers or electronic components, usually require the help of a specialist. For instructions on fixing leaking pipes, turn to page 43.

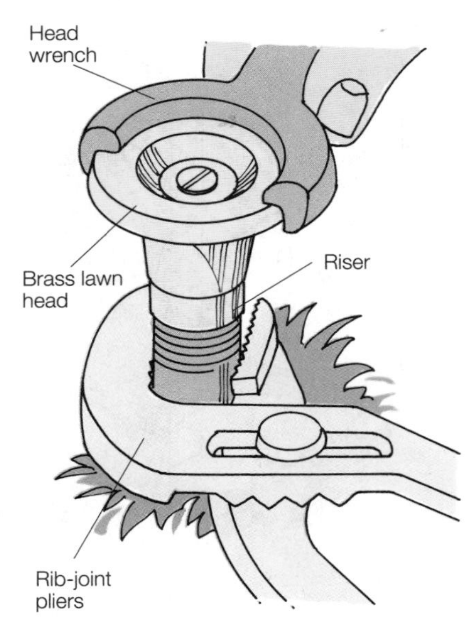

Drawing 10. To remove a damaged brass lawn head, hold the riser with rib-joint pliers and use a head wrench or a second pair of rib-joint pliers to turn the head.

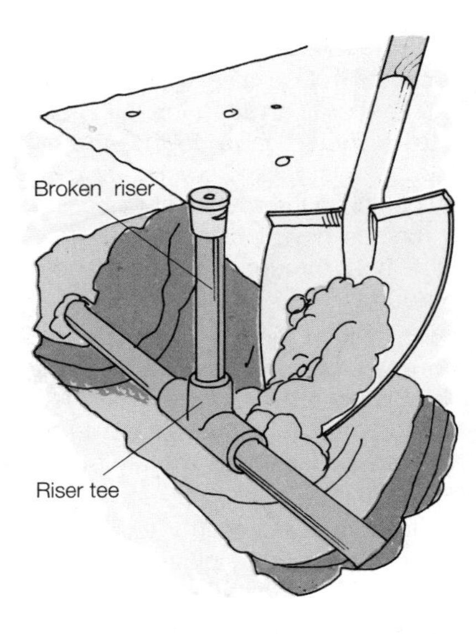

Drawing 11. To replace a broken riser, first dig down to the tee or elbow fitting, taking care not to damage the pipe. Clear soil away from the fitting; carefully unscrew the riser.

Replacing sprinkler heads & risers

If water streams or gushes from heads or risers (the vertical pipes below the heads) when you turn on the system, a head or riser may be loose, damaged, or broken.

If the problem is simply a loose head or nozzle, tighten it (be sure to point a directional head in the right direction). Instructions for fixing damaged or broken heads and risers follow. As you work, be careful to avoid getting dirt into the pipes, as it can clog the system.

Repairing damaged heads. The first step is to remove the head. Some heads are easily removed: with a shrub head, just unscrew the head counterclockwise; if it's too tight to turn by hand, use an adjustable-end wrench while holding the riser in place with a pipe wrench or rib-joint pliers.

To remove brass lawn heads, you may need a head wrench (see **Drawing 10**). Purchase one that's the same brand as your sprinklers or buy an adjustable universal head wrench. Old brass heads are sometimes difficult to remove. For more leverage, slip a pipe over the handle of the head wrench.

When you unscrew the head, try not to remove the threaded nipple riser that sits underground between the head and the water supply line. If you sense the riser is turning as you unscrew the head, remove enough soil so you can hold the riser with pliers or a pipe wrench. If the head has rusted onto the riser and you can't separate them, you'll have to replace both parts.

Replace the damaged head with the same type. If the new head sprays unevenly, remove and clean it; then flush the line to get rid of any dirt that might be clogging the system.

Fixing broken risers. If a riser is difficult to extract, carefully dig down to the threaded tee or elbow into which it's screwed (see **Drawing 11**), clearing away the surrounding soil. Unscrew the broken riser from the line. Wrap the threads of a replacement with fluorocarbon (pipe-wrap) tape and screw it in place.

Sometimes, a plastic riser will break off inside the threaded tee or elbow in the water pipe, leaving a piece that's difficult to remove. Use a stub wrench to remove a plastic riser piece, or try to unscrew the piece with a chisel, wedging it against one edge of the piece and twisting in a counterclockwise direction. Be careful not to damage the threads of the tee or elbow. Attach a new riser.

If you suspect that some dirt may have fallen into the pipe while you were working, remove all the heads on the circuit and turn on the water briefly to flush the pipes.

Unclogging sprinkler heads

When a sprinkler is clogged, it will usually force water out at odd angles, or the spray will be greatly reduced. Brass sprinkler heads clog frequently, since they don't have filters. Any head can become clogged if soil, mineral deposits, insects, or debris collects in the slit or holes where water emerges.

Several times during the season, check heads for uneven spray patterns while the system is on. Clean slits or holes with a knife or a piece of small, stiff wire. If the spray is weak on several heads near one another, you may have too many sprinklers on the line, or the water pressure may be too low to lift them (brass pop-up heads, which need high water pressure, are especially susceptible). You may want to replace the brass heads with low-pressure plastic heads.

Flood and stream bubbler heads have small openings that can't be cleared with a knife blade. Try cleaning flood bubblers by opening the adjusting screw to full flow. If this doesn't work, unscrew the heads and flush them under running water. If plastic heads are spraying erratically, unscrew the cap or nozzle and clean the filters.

With pop-up brass heads, debris can collect around the wiper seal on the stem, preventing the head from sealing; water will then squirt or bubble out from the base, and the spray will be weak. Clean around the seal.

Adjusting sprinkler throw, pattern & direction

Because water pressure varies during the day, it's important to inspect your system at the time it's normally in use. Turn the system on and look at all the sprinkler heads in your system, checking them for the correct throw, pattern, and direction. Make sure that the spray from each head isn't blocked by tall grass or plants, and keep grass clipped low around heads.

Old lawns often gain several inches of height from built-up thatch. If any lawn or shrub head sits so low that the spray is blocked, change the head to a pop-up type, replace the riser with a longer one, or add a coupling and another riser (see **Drawing 12**).

Sprinkler throw. It's important for complete coverage that the spray patterns of your sprinklers overlap. When overlap isn't sufficient, your lawn may develop brown patches, and plants may wither and die. Don't try to compensate by running the sprinklers longer than necessary; this wastes water. Instead, adjust or change your heads.

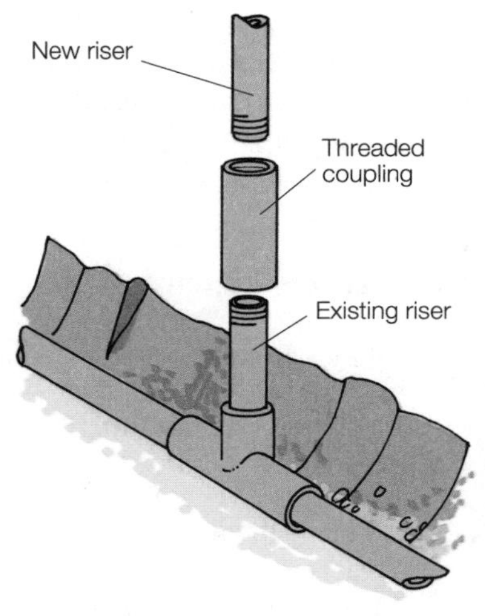

Drawing 12. To raise a sprinkler head, add a coupling and another riser to the existing one.

Most pop-up and spray heads have a small screw on top that allows you to adjust the flow of water (see **Drawing 13**). To increase the radius, use a small screwdriver and turn the screw counterclockwise. If the screw is already fully open, try increasing the flow at the valve, or replace the head with one that has a larger radius.

When a sprinkler is overshooting an area, reduce the flow of water by turning the screw clockwise. But don't try to reduce the radius by more than 25 percent (from fully open). Instead, replace the head with a lower-flow sprinkler model.

Sprinkler patterns. The pattern of the head is identified on the head itself by letters ("H" for half, "Q" for quarter, and so on) or numbers (for example, "¼" for a quarter-circle spray pattern). Plastic heads may also have score marks on them.

To prevent lawn browning and plant die-back, you need to have the correct spray pattern for the area you're irrigating. If a quarter-circle sprinkler head isn't fully covering the area around it, for example, try changing it to a half-circle head (with pop-up brass and plastic sprinkler heads, you only have to change the nozzle, not replace the entire head). Rotary heads have adjustable ring retainer clips that lock in the head sweep (see **Drawing 13**).

Direction of throw. Check the heads along the perimeter of the lawn and planting beds to make sure they're aimed in the right direction. If any of them are spraying off center, gently turn the heads or nozzles, or adjust the ring retainer clips on rotary sprinklers, until they cover the desired area.

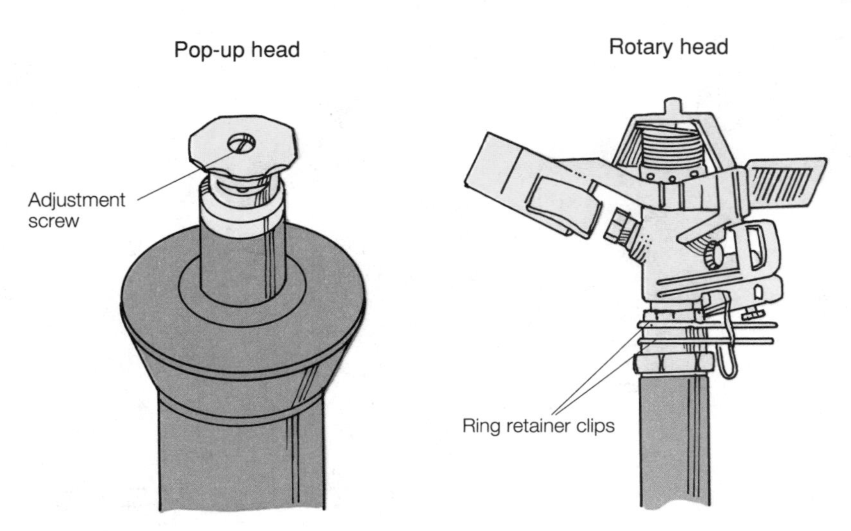

Drawing 13. To adjust a pop-up sprinkler head's distance of throw, use a screwdriver to turn the small screw on top. A rotary head has ring retainer clips that are used to set the sweep of the head.

Installing Drip Irrigation

Drip irrigation is the most practical, efficient way to water plants. In a typical system, water travels through lengths of polyethylene tubing; emitters attached to the tubing deliver water where you want it—near plant roots—in a gradual flow adjusted for the plant's water requirements. Drip systems eliminate water runoff and reduce water lost through evaporation and overspray by up to 70 percent.

Drip is especially useful in special planting situations. You can install inline emitters for plants growing between paving, create a multicycle mist system for frequent watering of shallow-rooted plants, make a ring of in-line emitters to water large containers, and run tubing along eaves to supply hanging pots. Used in conjunction with an automatic controller, or timer, such a system offers both flexibility and excellent control.

A complete drip irrigation system usually includes a range of components (see **Drawing 14**), designed to deliver water to all the plants in your garden. Most components are sturdy and hassle-free. Installing such a system is relatively easy, even if the area you're covering is extensive; for planning help, turn to pages 92–93.

Following is a rundown of typical components and how they operate.

Water-delivery components

Though they all deliver water to plants, emitters, misters, mini-sprays, and mini-sprinklers (see **Drawing 15**) play particular roles in a garden watering scheme. Knowing the differences will help you design the system that's best for your garden. For help in choosing the right emitters for your soil type and the plants you're watering, see the chart on page 100.

Emitters. Drip emitters deliver water from the main distribution tubing to plants. They're best for watering individual plants, since water placement is more precise than with sprays. Usually, emitters can be completely hidden from view.

Most emitters have barbed ends that snap into ½ or ⅜-inch polyethylene tubing, or push into the ends of ¼-inch microtubing. Some are barbed on both ends, so you can create a chain of tubing and in-line emitters. Others come preinstalled in tubing.

Emitters are color-coded for flow rate; red, green, brown, blue, pink, or black may signify rates of ¼, ½, 1, or 2 gph (not all manufacturers use the same colors to identify the same rates).

Multioutlet emitters have up to 12 outlets per head.

- *Diaphragm-type emitters* have an interior diaphragm that opens or closes to control flow as pressure changes. They're generally self-flushing and automatically compensate for varying water pressure.

  Diaphragm emitters are best for hilly terrain, slopes, and systems using long lines of emitters. Don't use them if your pressure at the water source is lower than 5 psi (pounds per square inch).

- *Turbulent-flow emitters* have twisting pathways that reduce pressure by creating turbulence; this also makes them partially pressure-com-

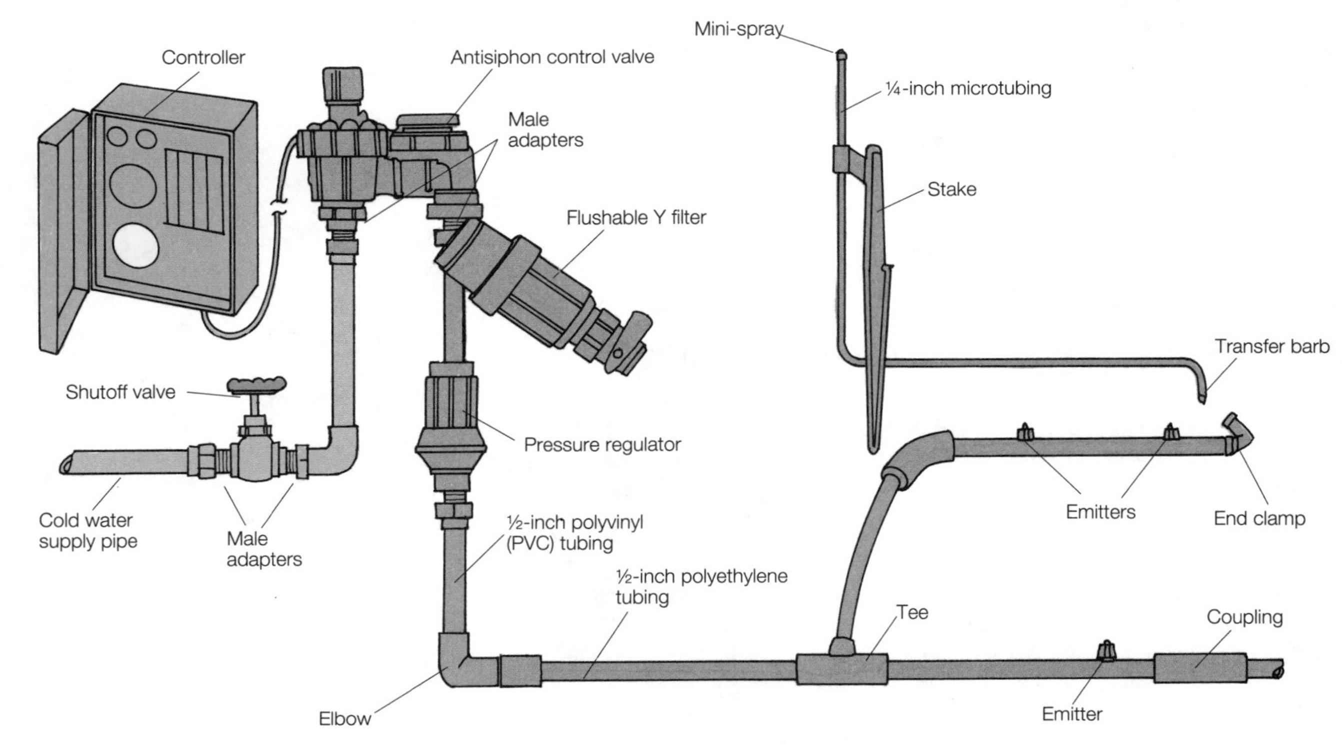

Drawing 14. Typical drip system components

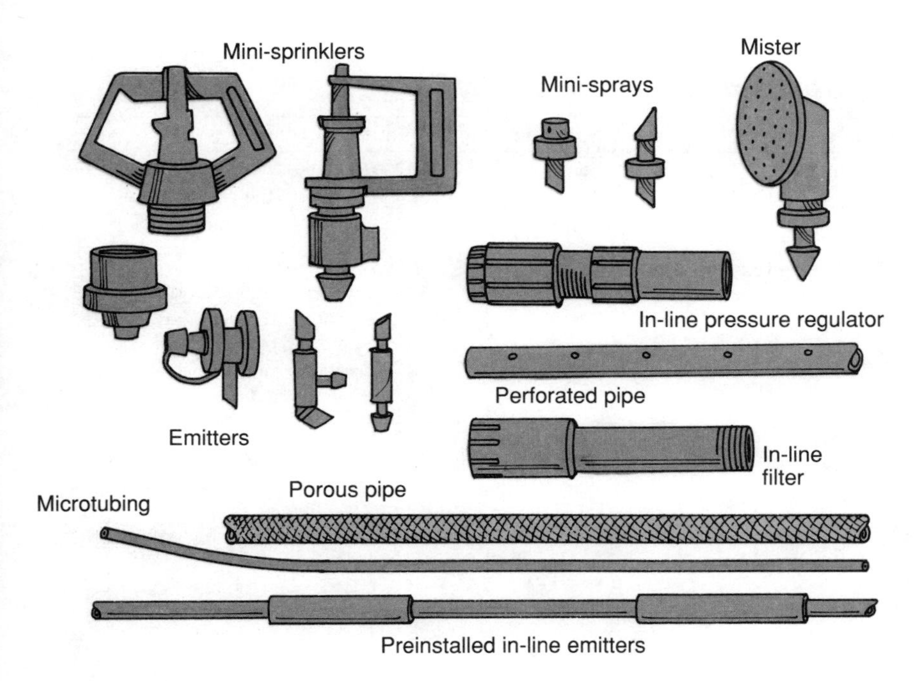

Drawing 15. Detail of drip components

pensating. The wide channels pass debris, so this type is less likely to clog, making these emitters especially useful where water quality is poor.

- *Vortex-type emitters* spin water in interior chambers to lower the pressure where the water exits. Water that's high in calcium tends to clog this type of emitter.

Misters, mini-sprays, and mini-sprinklers. Inserted into tubing or run off microtubing and mounted on stakes, these spread water over wider areas than do emitters but still operate at low flow rates and low pressure. They're particularly useful for closely spaced ground covers, flowers, and vegetables. As plants grow taller, they may block the spray, but you can rearrange sprayers or add new ones so plants get sufficient water.

All three types come in a range of flow rates. Higher water pressure increases the flow and gives wider coverage.

- *Misters* deliver a very fine spray. They're commonly used for ferns and other plants that need high humidity and frequent watering.
- *Mini-sprays* come in various spray patterns, including rectangular for strips and 90°, 180°, 300°, and 360° circular patterns that are good for irregular shapes as well as for open spaces.
- *Mini-sprinklers,* or spinners, emit large droplets that are less affected by wind than droplets from misters or mini-sprays. Their wide, full-circle patterns—from 10 to 30 feet—are useful for large areas.

Tubing

The standard way to distribute water in a drip system is through ½ or ⅜-inch polyethylene tubing attached with plastic fittings and laid on the surface of the soil, where the tubing can be obscured by a mulch.

Flexible and easy to cut, the tubing is connected without glue. Both inside and outside diameters can vary widely, so it's a good idea to keep a sample of the tubing you use on hand in case you need to add to the system later on.

Use ¼-inch microtubing (often called spaghetti tubing) to connect mini-sprays and mini-sprinklers to the distribution lines. Because it's easier to conceal than larger tubing, microtubing is also good for distributing water to containers and hanging baskets on decks and patios. Don't choose it for general use—it's fragile and too easily knocked away by rakes or animals.

Continuous-flow pipe

When you need to water a row of plants, continuous-flow pipe, which emits a continuous band of water, is a quick and easy solution because it eliminates the need for emitters. All types should be used with a pressure regulator and filter; the porous kind works best at 5 to 10 psi.

- *Preinstalled in-line emitters* give a continuous band of water. This dripline tubing is sturdy, clog-resistant, and generally trouble-free. It comes in 100-foot rolls.
- *Perforated pipe* is good for flower and vegetable beds, but not for permanent installation, because the holes have a tendency to clog.
- *Porous pipe* (ooze tubing) is best for underground installations, as in a vegetable bed, where the tubing stays moist. If the pipe dries out between watering cycles, calcium may build up inside and clog the pores. It's available in 100-foot rolls.

Supply controls

Controllers, valves, filters, and more specialized components automate your system, making it more versatile and efficient.

Controllers. The heart of the system, a controller, or timer, is an electronic device that automatically regulates the operation of each system connected to it. For all but the smallest system, choose a multiprogram automatic controller that allows you to set up watering frequency and duration. If you use two or more separate control valves, you can water on different schedules automatically.

Controllers that can be scheduled in hours, rather than just minutes, are the most versatile, since they run long enough to deep-water trees and shrubs. A less-expensive option is a controller that can repeat its cycle several times during the day.

. . . Installing Drip Irrigation

Valves. All systems need a control valve combined with an antisiphon device. For help in choosing the suitable one for your system, call your water department. Valves designed for low-flow shutoff (½ gpm or lower) are good for small systems.

One helpful device is an automatic rain shutoff valve. Mounted out in the open, on a fence or near your house, it measures rainfall and automatically shuts off your system when water reaches a certain (often adjustable) level. When the water evaporates, the system turns back on.

The simplest such devices rely on natural evaporation; mount them in direct sun. More sophisticated types have a heating mechanism that speeds evaporation, useful where humidity is high or brief summer storms are a common occurrence.

Filters. Most household water is clean, but sediment can get into the line during flushing of city water pipes or from old galvanized pipes in your house. That's why your drip system needs a good-quality 150 to 200-mesh flushable Y filter, one that uses fiberglass or stainless steel screens to filter out sediment that might clog the emitters. Install the filter just below the control valve.

Pressure regulators. Most low-volume systems are designed to run best at water pressures between 20 and 30 psi. However, household lines generally range from 50 to 100 psi, with some areas as high as 300 psi. To compensate, you need to install a pressure regulator (usually one for each main line) between the filter and the line. This device reduces the pressure to a rate that won't blow the system.

Fertilizer injectors. Spreading fertilizer over the soil is not an effective way to feed plants that are watered by drip irrigation. A fertilizer injector installed between the control valve and the pressure regulator puts fertilizer directly into the plants' water supply. Some injectors work by dissolving tablets, others by injecting liquid or soluble dry nutrients into the flowing water. Some injectors are integral with filters.

Which emitters are right for your system?

Plants	Soil type	Emitters
Low shrubs	Sandy soil	One 2-gph emitter next to plant
	Loam	One 1-gph emitter next to plant
	Clay soil	One ½-gph emitter next to plant
Medium-size to large shrubs	Sandy soil	Two or three 2-gph emitters placed evenly around plant
	Loam	Two or three 1-gph emitters placed evenly around plant
	Clay soil	Two or three ½-gph emitters placed evenly around plant
Small trees (to 8-foot-wide canopy)	Sandy soil	Three to six 1-gph emitters or two or three 2-gph emitters, installed on a J-loop or on two lines set on opposite sides of trunk
	Loam	Two or three 1-gph emitters, installed as above
	Clay soil	Two or three ½-gph emitters, installed as above
Larger trees (10 to 15-foot diameter)	Sandy soil	Four to ten 2-gph emitters, installed on a J-loop or on two lines set on opposite sides of trunk
	Loam	Four to ten 1-gph or three to six 2-gph emitters, installed as above
	Clay soil	Four to ten ½-gph or three to six 1-gph emitters, installed as above
Ground covers spaced at least 2 feet apart	Sandy soil or loam	One 1-gph emitter at rootball
	Clay soil	One ½-gph emitter at rootball
Closer ground covers with less distinct root zones	Any soil	Overlapping mini-sprays or mini-sprinklers (or see below)
Beds of flowers, ground covers, and vegetables	Sandy soil	Several 2-gph emitters spaced about a foot apart in a row
	Loam	Several 1-gph emitters spaced about 1½ feet apart in a row
	Clay soil	Several ½-gph emitters spaced about 1½ feet apart in a row
Container plants	Potting soil	One or more ½ or 1-gph emitters, depending on pot size

Emitter flow rate	Amount & pattern of coverage		
	Sandy soil	Loam	Clay soil
½ gph	1 sq. ft.	5 sq. ft.	11 sq. ft.
1 gph	5 sq. ft.	11 sq. ft.	18 sq. ft.
2 gph	11 sq. ft.	18 sq. ft.	31 sq. ft.

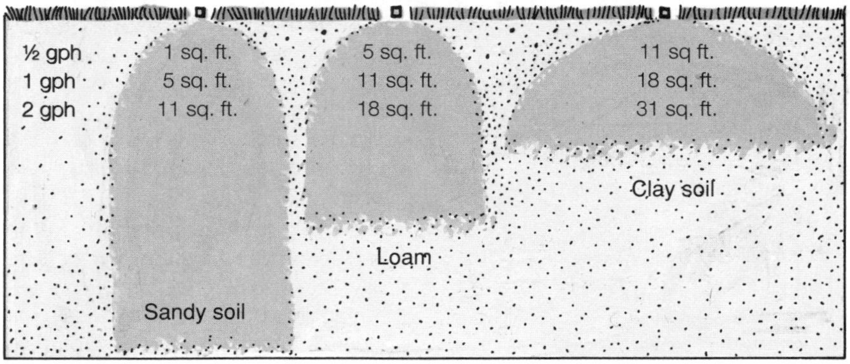

A gallon of water in a drip system moves differently through different kinds of soil. The numbers for each give the maximum horizontal coverage at different emitter flow rates (expressed in gallons per hour). The shading shows the vertical wetting patterns.

Converting Sprinklers to Drip Systems

Though automatic sprinkler systems take the tedium out of watering, most are overly generous with water. Like rain, they blanket areas, watering plants, weeds, and unplanted soil equally and offering little selective control. An effective way to minimize water waste where you have garden plantings is to convert your sprinklers to a drip irrigation system.

Hooking up a drip system to existing sprinkler piping is a relatively easy task. You first position a water filter and pressure regulator; then you adapt one or more risers for drip tubing connections and cap the rest; finally, you lay down tubing and position emitters as for drip irrigation (for details, see pages 98–100).

You'll need to convert each control valve completely from one system to the other—drip and regular sprinklers don't operate properly on the same circuit.

The parts you'll need are available in irrigation supply and hardware stores.

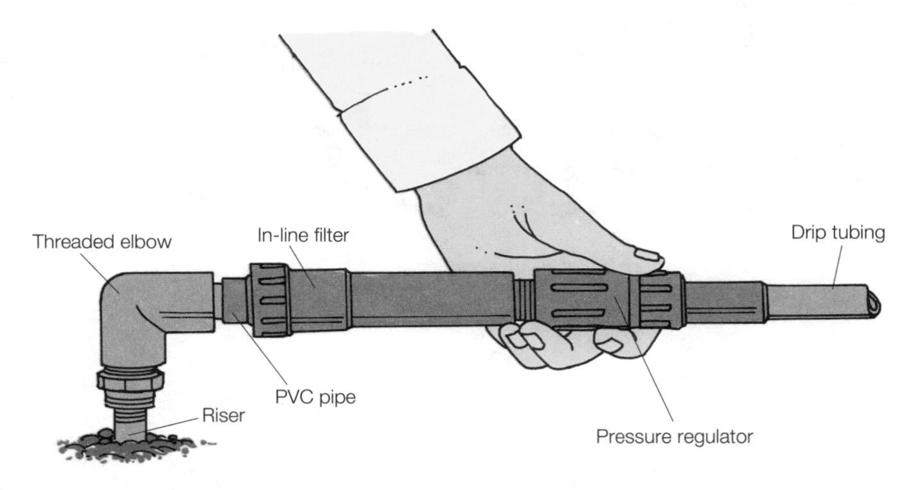

Drawing 16. Adapt a riser to drip irrigation by installing an in-line filter and pressure regulator.

Locating a filter & pressure regulator

To convert to drip, the water must be filtered to prevent clogs, and the in-line pressure must be reduced to 20 to 30 psi (pounds per square inch). If you're converting just one sprinkler head to drip, you can simply add a filter and pressure regulator to the line at a riser (see **Drawing 16**). If you're converting more than one sprinkler, you attach the filter and pressure regulator at the valve.

Installing at a riser. Since drip lines can run 100 feet in any direction from one sprinkler riser, you may only need to convert one riser. To adapt a riser for a drip line, screw a threaded PVC elbow or tee fitting onto the riser and add a short length of threaded PVC pipe. Add the filter and pressure regulator; then attach drip irrigation tubing to the pressure regulator and extend it as needed. Remove the other sprinkler heads on the line and cap the risers with threaded caps.

Keep the filter and pressure regulator aboveground so the filter can be cleaned.

Installing at the valve. Installing a filter and pressure regulator at the valve allows you to convert all the sprinkler heads on that circuit to drip.

The fittings and adapters you'll need depend on your system. Take a drawing or photo of your valve setup (with measurements) to the store and do a trial assembly of the parts.

To install them, you'll need to remove a section of pipe below the control valve. Dig down just beyond where the line makes a 90° turn toward the sprinklers and remove a section of the vertical pipe. Replace it with the filter and pressure regulator. Adapt those sprinkler heads you want to convert and cap the rest.

Screwing high-pressure emitters onto risers

High-pressure emitters (see **Drawing 17**) are useful where you have clusters of plants around each riser rather than spread out over a large area. These emitters are installed directly onto risers.

Each emitter has from four to twelve outlets and a built-in pressure regulator and filter. You'll need to add more filters only if your water pressure is higher than 100 psi, you irrigate with well water, or you're using laser-drilled tubing.

If your system is galvanized pipe rather than PVC, wrap the riser three times with fluorocarbon (pipe-wrap) tape and hand-tighten the emitter onto it so the plastic doesn't crack; or change to a PVC riser. Attach microtubing for shrubs and trees or laser-drilled tubing for closely spaced plants.

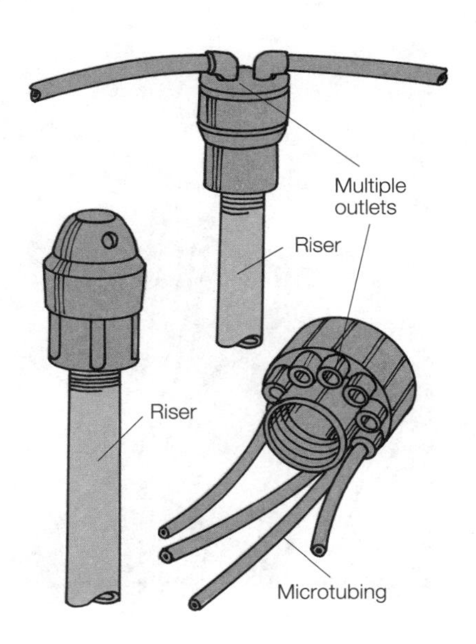

Drawing 17. High-pressure emitter heads screw directly onto risers; on some, microtubing runs from the outlets.

Water-saving Strategies

Conserving Water Inside & Out

Did you know that a leaky faucet in your house can waste as much as 24,000 gallons of water every year, or that recent studies show that homeowners watering their lawns consistently apply at least twice as much water as their lawns actually require? In some areas, the need for conserving water is critical. But no matter where you live, wasteful water practices can be both expensive and ecologically unsound.

Using water wisely

The first step in conserving water is to understand how much water you and your family use for various activities. (To determine your water usage, see the chart below.) With that information, you can better plan and manage your use of water.

Inside a house, more water is used for flushing toilets than for any other activity. Next on the list are showers and baths, and then washing machines and dishwashers. Whenever possible, convert your water-wasting fixtures to water-conserving ones (for information, see pages 105–106).

Remember that your everyday practices are a key aspect of water conservation. Changing how, and for what, you use water can be your most effective water-saving strategy.

Saving water in the house

You can also significantly reduce water consumption in your home by adopting some or all of the following practices.

- *Don't let water run.* Turn it off while you're shaving, brushing your teeth, or doing dishes by hand. Prerinse dishes for the dishwasher in a pan, not under running water, or just scrape them with a utensil or wipe them with a napkin.
- *Don't use running water* to thaw frozen food.
- *Chill drinking water* by putting some in a container and storing it in the refrigerator.
- *Avoid using your garbage disposer* more than necessary; it requires lots of water to work properly. Instead, throw trimmings in the trash, or, better yet, put them in your compost pile.
- *Keep a basin in the kitchen sink.* Wash vegetables in it, cleaning them with a brush, or use the basin to catch water in which you've rinsed fruits, vegetables, and the like. Reserve cooking water and use it to water outdoor plants or to run the garbage disposer.
- *Take shallower baths.* Put the stopper in before you start to run the water: that first bit of cold water will quickly mix with warm to give you the desired temperature.
- *Take shorter showers.* As the shower water warms up, capture it in a bucket and then set the bucket at your feet while you shower. Pour a bucketful of water into the toilet bowl to flush it manually.
- *Flush the toilet only when necessary.* Never use it as a wastebasket or ashtray.
- *Attach a combination stream-spray aerator* to the bathroom sink faucet if the spout will accommodate one. Use the steady stream to quickly wet a toothbrush; for hand-washing, use the more efficient spray.
- *Run dishwashers and clothes washers fully loaded,* even if they have adjustable water levels. Use the shortest cycles possible on both machines. (For more on energy-saving dishwashers, see page 59.)

Saving water in the garden

One of the best ways to conserve water in your garden is to use drip irrigation

Where does your water go?

Indoors	Gallons used
Washing machine (standard)	35 to 70 per load
Toilet (standard)	3½ to 7 (or more) per flush
Toilet (ultra-low flush)	1.6 (or less) per flush
Toilet with silent leak	40 per day
Toilet that "runs"	5 per minute
Dishwasher (standard)	14 to 30 per load
Dishwasher (water-saver)	11 to 27 per load
Sink-washing dishes (tap running)	5 per minute
Sink-washing dishes (filled sink)	5 to 10 per wash
Garbage disposal	3 to 5 per minute
Faucet (kitchen)	5 per minute
Faucet (bathroom)	5 per minute
Faucet, leaky	⅓ per hour
Shower head (standard)	5 to 8 per minute
Shower head (low-flow)	1½ to 3 per minute
Bathtub	25 to 35 per bath

Outdoors	Gallons used
Lawn (20 by 40 feet)	2,000 to 4,200 per month
Sprinkler (standard)	½ to 4 per minute
Drip-irrigation emitter (one)	¼ to 2 per hour
Hose (½-inch diameter)	300 per hour
Hose (⅝-inch diameter)	500 per hour
Hose (¾-inch diameter)	600 per hour
Faucet with slow drip	350 per month
Faucet with fast drip	600 per month
Faucet with fast leak	2,000 per month
Washing car (hose running)	100 to 200 per 20 minutes
Washing car (with hose shutoff)	15 or more per 20 minutes
Uncovered pool (18 by 36 feet)	900 to 3,000 per month
Covered pool (18 by 36 feet)	90 to 300 per month

. . . Conserving Water Inside & Out

for your plants. If you already have a sprinkler system, you can convert some of your heads to drip. (For more information on watering, see the chapter that begins on page 90.)

Here are some additional ways to conserve water outside your house.

- *Adjust watering schedules.* Let the weather be your guide for watering frequency.
- *Build watering basins.* They direct water right to the roots of shrubs and small trees.
- *Control runoff on slopes.* Put headers or basins downslope from plants.
- *Irrigate early or late in the day.* In the morning and evening, the air is still and evaporation is minimal.
- *Maintain drip systems.* Periodically check them for clogged or broken tubing or emitters.
- *Maintain sprinklers.* Inspect your system at least once a month during the watering season. Clean clogged sprinkler heads, replace broken sprinklers or risers, and adjust heads so their spray doesn't wet any paved areas.
- *Mulch plantings.* Materials such as ground bark, compost, and leaf mold help to keep the soil cooler, reduce evaporation, and discourage the growth of weeds. Under large trees, let fallen leaves or needles accumulate as natural mulch.
- *Pull weeds or mow them.* Left to grow, weeds compete with ornamentals for limited water.
- *Put up shadecloth.* Protect tender plants, such as young Japanese maples, from hot sun by covering them with shadecloth.
- *Repair leaks.* Fix dripping outdoor faucets and bad hose connections as soon as they occur.
- *Shade strawberries.* Cover strawberries with shadecloth, but be sure to allow good ventilation. After harvest, withhold water from most of the plants; continue watering a few so you can divide and replant in autumn or winter.
- *Sweep driveways and paths.* Use a broom rather than a hose spray to clean off paving.
- *Water cane berries sparingly.* After harvest, especially in coastal climates, established blackberry plants can get through summer on no water; raspberry plants will survive on very little.
- *Time watering carefully.* Use your irrigation system's automated controller, changing it as necessary for different weather conditions. If your system is manual, set a kitchen timer to remind you when it's time to turn off the water.
- *Use soaker hoses.* Though they're less efficient than drip, soaker hoses are inexpensive and easy to use between rows of vegetables and around big trees.
- *Thin deciduous fruit.* Thin apples, peaches, and plums to 10 to 12 inches apart. Let the trees go dry after harvest; water only if the leaves begin to wilt.
- *Water efficiently when using a hose.* Equip the hose with a shutoff valve so you can turn off the water as you move from plant to plant.
- *Water roses sparingly.* After spring bloom, many established ones, especially old shrub and species types, can get by with little water. Don't deadhead; let hips develop to suppress plant growth.
- *Withhold fertilizer.* Don't feed trees or shrubs; fertilizer stimulates new growth, which increases the demand for water.

Detecting a running toilet

A toilet that runs with a steady flow can waste as much as 5 gallons of water a minute, which adds up to 7,200 gallons per day. Of course, you can hear the flow of a toilet that runs this heavily.

Harder to detect is a toilet that leaks slowly. Even an imperceptible toilet leak can waste as much as 40 gallons of water each day.

To check your toilet for a leak, remove the tank lid and add about 12 drops of colored dye to the water inside. (Blue food coloring will work, or ask your water company for dye tablets.) Then wait 5 minutes to see if the dye flows into the bowl. If it does, water is escaping either through the top of the overflow tube or past the tank stopper and valve seat.

A float ball that's riding too high lets water overfill the tank and spill down the overflow tube into the bowl. When the chain or lift wires between the flush handle and the stopper at the bottom of the tank get kinked or fouled, the stopper is prevented from sealing the valve seat. Or the stopper or valve seat may be worn and closing improperly (this is usually the problem if you can jiggle the handle to stop the leak).

To make the repairs yourself, turn to pages 32–37.

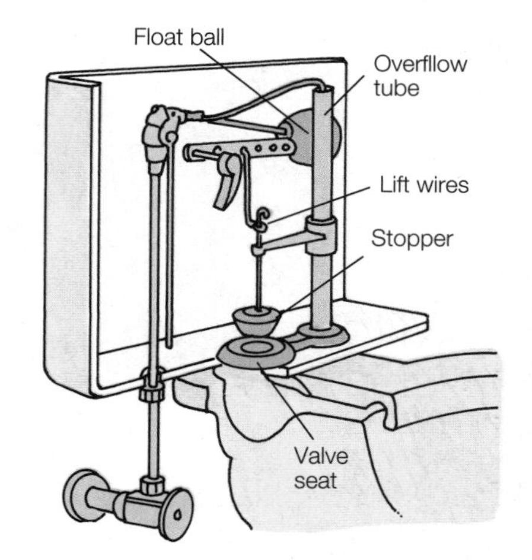

Drawing A. Components of a toilet

Water-conserving Fixtures

Retrofitting or changing some of the water-using fixtures inside your house to make them more water efficient can result in substantial water savings. Some changes, such as retrofitting toilets and replacing standard shower heads with low-flow heads, require little effort. Replacing a toilet is more complicated but still within the realm of the do-it-yourselfer.

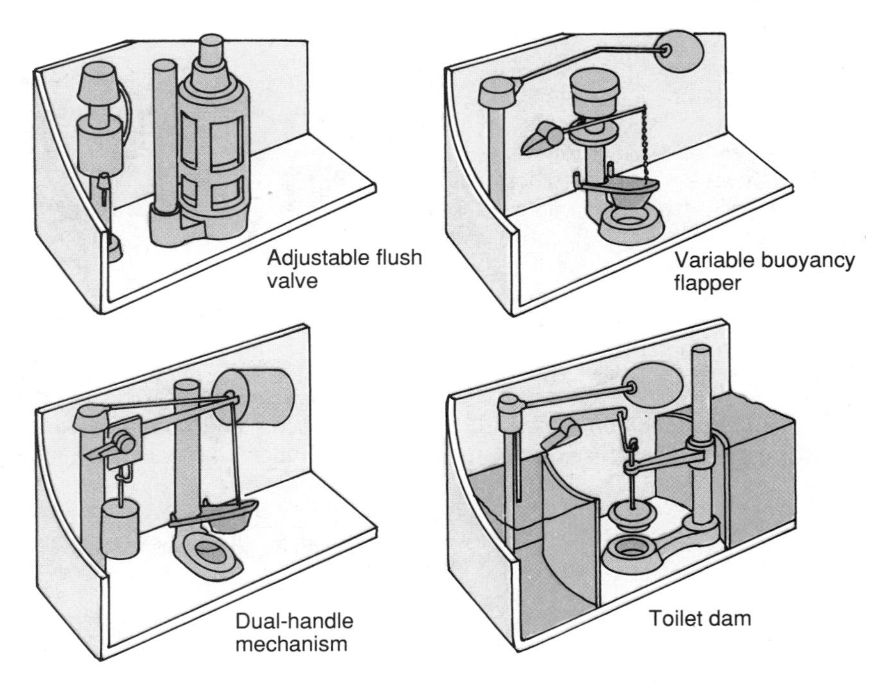

Drawing 1. Retrofit toilet devices include an adjustable flush valve, a variable buoyancy flapper, a dual-handle flush mechanism, and a toilet dam. Each reduces the amount of water used for flushing.

Retrofitting & replacing toilets

Because toilets consume more water than any other household appliance or fixture, they're the best place to begin upgrading for water conservation. If your existing toilets are water wasters—requiring more than about 3½ gallons per flush—you can either retrofit them with devices that cut back water usage or replace them with low-flush models.

All new toilets must meet the American National Standards Institute's (ANSI) stringent hydraulic performance standards, which cover bowl cleaning, removal of solids, and drain line carry over a distance of 40 feet. On new toilets, look for a stamp or sticker of approval; the number of gallons used per flush should be printed in the tank or on the packaging.

Before installing a new toilet in an older house, check the offset, the distance between the back wall studs and the center of the drain hub (measure to the hold-down nuts). New toilets are designed for a 12-inch offset. In older houses, the offset may be greater, pushing the toilet away from the wall.

For instructions on replacing a toilet, turn to pages 78–79.

Retrofitting your toilet. Installing any of several water-saving assemblies in your toilets (see **Drawing 1**) can make them more efficient. Most devices can be installed in just a few minutes; simply follow the manufacturer's instructions.

One device, an adjustable flush valve, replaces a conventional ball cock and stopper, or flapper. You adjust it to close once flushing is complete and the tank and bowl are filled to a given level. The adjustable valve prevents water from escaping down the flush valve while the tank is refilling. This device will typically save from ¾ to 1½ gallons of water per flush, depending on the adjustment and your toilet.

A related device, the variable buoyancy flapper, rides up and down on the overflow tube and closes the stopper before all of the water rushes from tank to bowl. The water that does rush through still moves with its original full force. When such a device is put into a 5-gallon toilet and adjusted to save 2 gallons, all 5 gallons of water are still moving downward once flushing starts, but the stopper will close while 2 gallons still remain in the tank.

Dual-handle flush mechanisms work in a similar way, but they give you a choice. You can still use the full flush with one handle, but you have a second handle that releases up to 75 percent less water—still enough force to carry away liquid waste.

Water-displacement devices, such as jugs of water and dams, reduce the amount of water that flushes from a toilet's tank. But because they displace the amount of water in the tank, they also reduce the force of the water rushing into the bowl, the action that makes a typical siphon-wash toilet work properly. The key is to size the displacement device to allow enough water for a proper flush.

A best bet: ULF toilets. A family of four can cut indoor water use by up to 20 percent simply by replacing existing toilets with new ultra-low-flush (ULF) ones. Extremely water efficient, ULF toilets use no more than 1.6 gallons per flush, compared with 5 to 7 gallons or more for older toilets. Some water districts even offer a rebate if you install a ULF toilet.

Most first-generation water-efficient toilets simply reduced the amount of water in each flush cycle without changing much else. Double-flushing and drain clogs were common with these "improved" versions.

The ULFs, however, represent new engineering as well as new design. Steeper bowl sides, shallower traps, smaller siphon outlets, and 5-gallon tanks that only release 1½ gallons of water per flush (but use all 5 gallons

. . . Water-conserving Fixtures

of force and pressure) make them perform better than many units that use twice the amount of water.

Other water-saving toilets. Among the other water-efficient toilets you can buy (check local codes first) are a gravity-flush one that uses as little as 3 liters (3.3 quarts) of water, a 1-quart vacuum-assist flush model, and a "foam-flush" toilet that uses only 1 cup of water (a detergent foam generator eliminates friction inside the bowl).

Saving water in the shower

Another way to reduce water consumption significantly is to limit the amount of water that flows through your shower head.

Low-flow shower heads and flow restrictors. Installing a low-flow shower head or flow restrictor (see **Drawing 2**) is an easy way to cut down water use. Depending on head design and water pressure, such devices can reduce maximum flow to less than 2½ gallons per minute, compared with 5 to 8 gallons per minute put out by standard fixtures. A flow restrictor is simply a washer or insert that you install between the inlet pipe and shower head to restrict the amount of water that passes through the shower head.

The cost of low-flow heads and flow restrictors is minimal. But be aware that less expensive models deliver such fine water droplets that they won't wet your body very quickly—and may even feel a little cool by the time they get down to your knees.

Quality low-flow shower heads are designed to give you a satisfying shower even though they sharply reduce the amount of water required. Inside these heads, water travels through special chambers and orifices that size water droplets, tightly focus the stream, and—with some types—mix air with the water to create a forceful, turbulent spray.

Low-flow heads come in two types: stationary fixtures (typical shower heads) and hand-held models. Hand-held ones, like those described on page 75, are more versatile and use less water than stationary heads because you direct the flow. Most hang on a wall-mounted hook or bar.

Each model offers different features. Look for adjustable spray settings and a head with a shutoff that reduces flow to a trickle while you soap up (see below).

To replace a shower head, unscrew the old head by hand or with an adjustable wrench. If necessary, hold the inlet pipe in place with a pair of locking pliers (wrap the jaws with a rag to keep from scratching the finish). Follow the manufacturer's directions for installation of the new head.

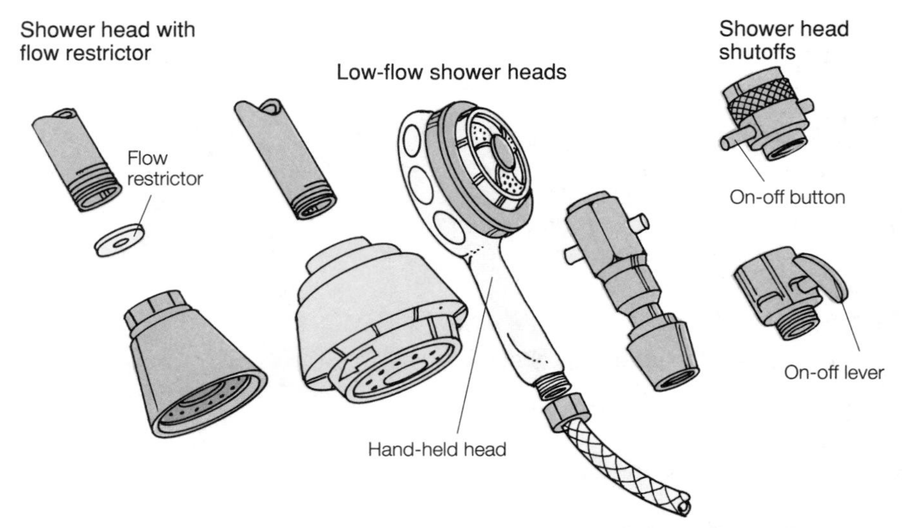

Drawing 2. Flow restrictor, low-flow shower heads, and shower head shutoffs

Shower head shutoffs. Also consider a valve near the shower head that lets you shut off the water while you're soaping up or shampooing (see **Drawing 2**). These valves, built into many of today's shower heads, dribble when closed so water in the pipe stays at the selected temperature. Generally, your water will be off for about half your shower time, so water savings will be proportional.

Kitchen water savers

Faucet shutoffs are available for kitchen sinks (see **Drawing 3**). Though it seems easy enough to simply turn off the water as you're washing vegetables or dishes, an at-the-tap valve makes the task virtually effortless, so you'll be much more apt to use it. Some are combined with an aerator, which mixes air into water, reducing the amount of water needed by up to 60 percent. Others are threaded so you can reconnect your old aerator.

An instant hot-water device (see page 82) will give you 190° to 200° water on demand. No water is wasted for warm-up.

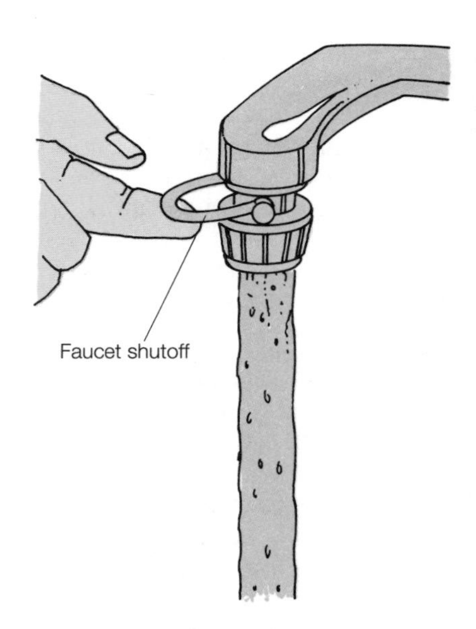

Drawing 3. Faucet shutoff for a kitchen sink offers convenient and immediate water control.

Underground Leaks

Leaks in the underground pipes bringing water to your house can result in losses that make your water bills read as though you're squandering water, even when you're doing your best to conserve. Such leaks can easily go unnoticed when camouflaged by lawn sprinklers, heavy irrigation, or plantings.

Here's how to find underground leaks outside your house. (Instructions for dealing with leaks in interior pipes appear on page 43.)

Is there a leak?

It's simple enough to find out if you're "consuming" water when you haven't even turned on a spigot. Go around the house and make sure everything is off: appliances should be through cycling (turn off the icemaker), and faucets should be shut. Then open the cap of your water meter to expose the valve gauge and put a mark on the rim where the dial needle is pointing. (If you have several dials, mark the one indicating 1-cubic-foot increments, as shown on page 10.)

Don't run any water for 30 minutes. Then check the meter. If the indicator has moved from your mark, you have a leak. (The meter may also have a little triangle that spins to show flow; its movement is further evidence of leakage.)

Determining water loss

By keeping track of both the time and the dial's movement, you can figure the amount of water loss. If you waited 30 minutes between checks, multiply the incremental change by 2 to get an hourly total. The gauge is calibrated in cubic feet; 1 cubic foot is about 7½ gallons. So a 2-cubic-foot change in 30 minutes would be 2 x 2 x 7½: a loss of 30 gallons per hour.

Locating the leak

Walk around the outside of your house, checking where you know pipes run. Look for any wet spots or sinkholes in your lawn; check paving and walls for cracks.

If you have a swimming pool, be aware that its pipes can leak, too. If the water level in your pool drops much more than ¼ inch per day under normal, calm conditions—or more than ⅓ inch in a desert climate—you may have a leak (a covered pool should show virtually no loss). Test by putting a piece of tape on the tile at water level; check again in 24 hours.

If you see no evidence of water loss, consider calling in an expert. In the past, the way to find a leak was for a plumber to make an educated guess, followed by trial-and-error digging. Today, electronic leak-detection devices can identify a leak's location to within a few inches, even through several feet of concrete; they can also determine the depth of the leak. To find a leak-detection service, look in the Yellow Pages under "Pipe & Leak Locating."

Most professional leak-detection services will fix a leak if they find one (if you want to do your own repairs, see page 43). They'll also test the line to see if there are any additional water losses that were obscured by the main leak. As a bonus, the leak-detection service will allow you to map most of your exterior plumbing lines for future reference.

Basic water-saver: a pool cover

Unless a swimming pool is covered, each summer month hundreds of gallons of water will literally vanish into thin air. Depending on climate and location, an 18 by 36-foot pool can lose from 900 to 3,000 gallons of water per month. Considered another way, an uncovered pool with a 20,000-gallon capacity could lose its total volume of water to evaporation in a typical one-year cycle. For a family of four, that represents 83 days of normal living, figuring 60 gallons of water per person per day. A pool cover can prevent more than 90 percent of this potential loss.

The two basic styles of cover are those that float on the surface and those that are mounted to the pool's edge or decking. The floating, or "solar blanket," styles are the least expensive. However, they're potentially the most dangerous. Because a free-floating cover offers no support, a child or pet could slip under the water and be unable to push the cover away.

The most familiar and least expensive floating covers are "bubble pack" covers, made of 12-mil plastic with an ultraviolet-light inhibitor. Another type of floating cover is made of closed-cell foam covered with woven polyvinyl.

Of the mounted covers, the best ones are made of vinyl with a dacron mesh, and contain algicides and ultraviolet-light inhibitors. The safest ones completely cover the pool and will support the weight of several adults.

Some mounted covers are manually operated—you pull them up out of the water or roll them up on rollers mounted at poolside. These covers can be cut to fit irregular shapes.

Automatic pool covers require tracks with built-in pulley systems to guide them into position. Though the tracks can be built unobtrusively into the edge of a pool during construction, for retrofits they usually must be screwed to the decking. Because the covers run in parallel tracks, they can only be rectangular in shape.

Gray Water Systems

Gray water refers to household water recycled from showers, bathtubs, bathroom sinks, and washing machines. Gray water collected from all these sources can supplement your garden's water supply by at least 100 gallons per person per week.

Be aware that not all household water can be reused. For example, much of the water from dishwashers and kitchen sinks (except rinse water) is contaminated with grease and food particles. Toilet waste water, called black water, should never be collected.

Precautions for using gray water

Even the use of "clean" gray water raises safety concerns, the primary one stemming from gray water's highly variable quality and possible contamination by bacteria, viruses, and parasites. Individuals can be infected by swallowing or inhaling those organisms.

Though health officials have frowned on the use of gray water in the past, decreasing water supplies have caused some communities to allow its use for specific purposes. For the rules about gray water in your area, call your building, health, or water department. Still, be advised: Most building codes prohibit altering household plumbing to collect gray water without a permit.

Minimizing health risks. The health risks from gray water are minimal if you collect and apply it properly. Never collect wash water contaminated by soiled diapers or by people with infectious diseases, or used in poultry or wild game preparation. Also, avoid collecting greasy, soapy water full of suspended solids.

If you collect gray water, don't wash strong chemicals down the sink. Such chemicals include products for opening clogged drains and products containing boron, chlorine, or sodium. Also, don't collect water that has run through a water softener. Of the standard laundry soaps, liquid detergents have less sodium than powdered ones, and biodegradable ones are usually the least harmful. Most plants do well on gray water and will not be affected by mild soaps or shampoos.

When using gray water, be sure to follow these basic precautions:

- *Always use gray water* the same day it's collected.
- *Don't let it puddle or stand* where it won't be absorbed quickly.
- *Don't spray or sprinkle it.*
- *Wear rubber gloves* whenever you handle gray water or equipment that comes in contact with it.

Where to use gray water in the garden. You can use gray water to irrigate fruit trees, ornamental trees and plants, flowers and ornamental ground

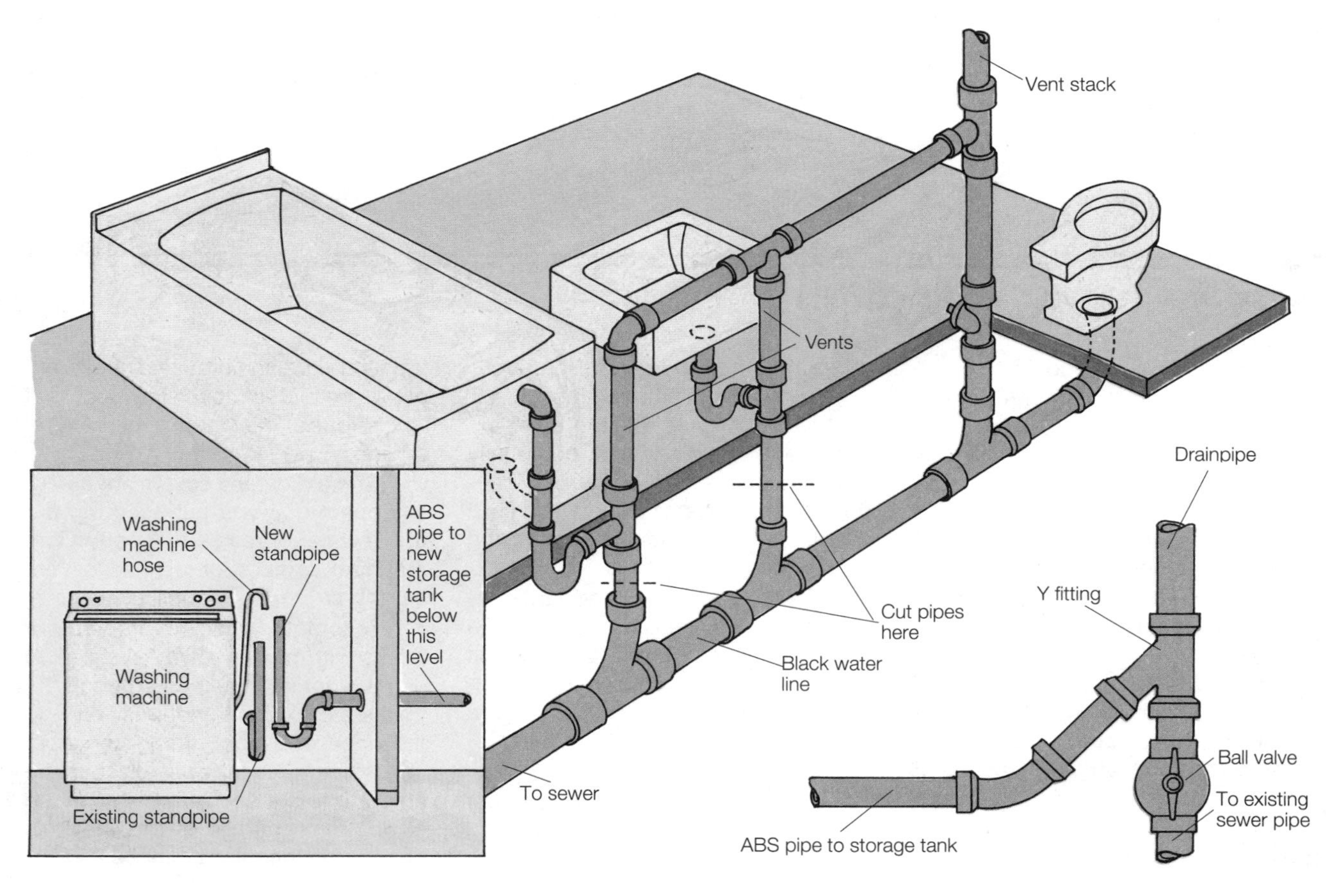

Drawing 4. To collect gray water, cut into the vertical drainlines of sinks, showers, and tubs; install Y fittings and ball valves as shown at right. New piping carries washing machine water (see inset) to the storage tank when desired.

covers, and lawns. Do not use gray water to irrigate vegetables or fruits you intend to eat without peeling or cooking. Exceptions are crops such as corn, beans, and tomatoes, where the edible parts are well above the ground. Don't use it for seedlings, container plants, or acid-loving and salt-sensitive plants.

Collecting gray water

A gray-water system installation usually involves connecting new pipes to the drainpipes of some plumbing fixtures, diverting the gray water through those pipes to a storage tank, where it's filtered and stored for up to a day, and then distributing the water. If this type of collection is legal in your area, you'll probably need a permit; washing machine plumbing may be excepted.

Making drain connections. Cut into drainpipes (see **Drawing 4**) from sinks, bathtubs, showers, and the washing machine. (To collect water from main plumbing lines on showers and many bathtubs, the pipes must be accessible from a crawlspace or basement.) Install Y fittings and ball valves to divert water either to your gray water system or to the standard drain. Use ABS (acrylonitrile-butadiene-styrene) drainpipes to carry the gray water directly to the storage tank.

Always provide for overflow into the sewer line in case the system backs up, and install a check valve between the tank and the sewer line to prevent sewage backup into the tank.

Making a tank. Tanks can be made from metal drums, plastic or fiberglass drums made for food, or 35 to 55-gallon plastic trash cans (see **Drawing 5**). Ideally, the tank should be located above the landscaping to be watered so water can be distributed by gravity.

To screen out hair and large particles in the gray water, make a shallow basket from ¼-inch hardware cloth and hang it inside the tank just below the rim. Provide handles so you can lift it out frequently for cleaning. Lining

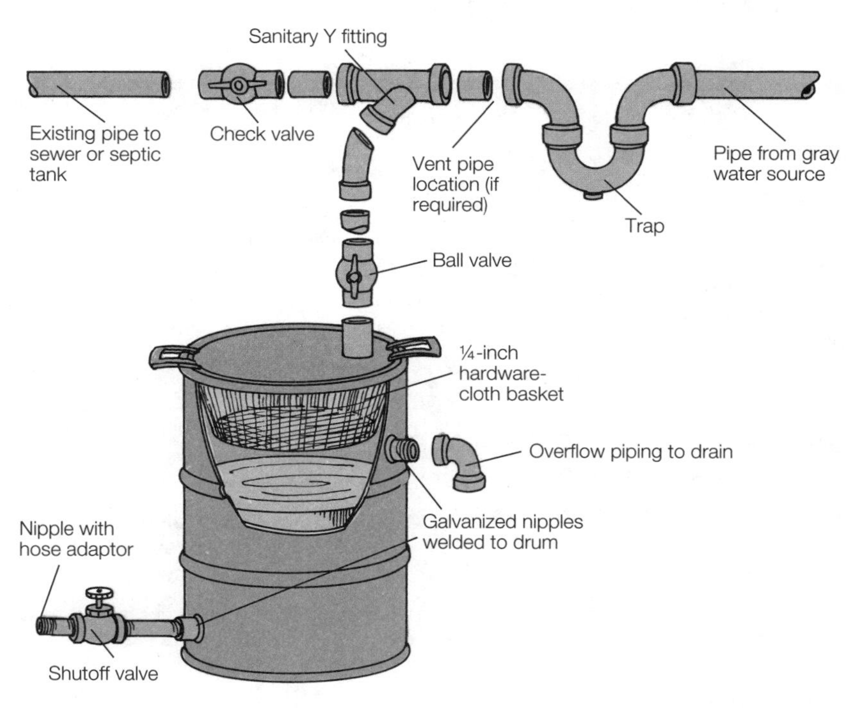

Drawing 5. Collection tank for a gray water system filters the water and stores it for up to a day.

the screen with a bit of straw makes it easier to clean.

Near the bottom of the tank, install a shutoff valve and a fitting that will accept a ¾-inch garden hose. For a plastic drum, drill a hole and install a tank nipple, a shutoff valve, and then the ¾-inch hose fitting. For a metal drum, have a welding shop install a ¾-inch metal pipe nipple. Add a nipple for overflow piping near the top

If the tank sits below grade or the garden slopes up above the outlet, install a sump pump. Add a check valve between the pump and the irrigation line to prevent backup.

Remember, too, that no matter how well the water is filtered, storage tanks need periodic cleaning. Use a hose to flush the tank; then drain the tank by opening the shutoff valve.

Distributing gray water

Health officials believe that the safest way to apply gray water is below the soil surface or under 4 inches of mulch. Below-surface applications eliminate human contact, and the soil filters out any harmful organisms.

To do this, you bury perforated pipe in a shallow trench filled with gravel. Buy 10-foot lengths of 3 or 4-inch rigid styrene drainpipe, with double rows of holes drilled along the length. Glue a 90° sweep fitting onto each end and add short lengths of pipe that will stick about 6 inches above the ground. Press a cap fitting onto each pipe (don't glue the fittings).

Shovel a few inches of 1 to 2-inch crushed rock into the trench, set the pipe in place, and level it. With the short pipe lengths pointing upward, bury the pipe beneath a few more inches of gravel; then top with soil.

Put in one of these 10-foot-long systems wherever you intend to deliver gray water. When you're ready to water, remove one cap, insert the hose that connects to the storage tank, and open the valve or turn on the pump. Simply move the hose from one system to the next to complete watering.

Don't try to connect a drip irrigation system to gray water distribution. Water for drip irrigation must be screened through fine filters that will clog frequently if gray water is used.

Plumber's Language

ABS (acrylonitrile-butadiene-styrene). Rigid plastic drainpipe.

Adapter. *See* Transition fitting

Aerator. Sievelike device on a spigot end that mixes air with water flow.

Air chamber (or air cushion or water hammer arrester). Device attached to supply pipes near outlets to prevent water hammer.

Air gap. Capped pipe that vents a dishwasher.

Angle valve. Valve with a 90-degree bend that eliminates need for an elbow.

Antisiphon valve. Valve installed on a supply line to prevent siphoning of contaminated water back into potable water supply system.

Auger. *See* Snake

Back venting. Vent looping up from a fixture and connecting to main soil stack or a secondary vent at a higher level.

Ball cock. Assembly inside a toilet tank that connects to water supply and controls flow of water into tank.

Balloon bag. Device that attaches to a garden hose and is used to loosen a blockage in a clogged drain.

Basin wrench. Tool designed to install or remove hard-to-reach locknuts holding a deck-mounted faucet onto a sink.

Bell and spigot cast iron pipe. *See* Hub

Bibb. *See* Hose bibb

Bleed. To drain a pipe of excess air by opening a valve at end of pipe.

Bonnet. Casing for wall-mounted bath and shower faucets; screws into faucet body behind wall.

Cap. Fitting with a solid end used for closing off a pipe.

Caulking. Material used to create a watertight seal.

Center-to-center. In mounting faucets: distance between centers of holes on a sink deck. In pipefitting: distance between centers of two consecutive fittings in a run.

Cleanout. Opening providing access to a drainline or trap under a sink; closed with a threaded plug.

Closet auger. Tool for clearing blockages in a toilet.

Closet bend. Drainpipe that joins to a toilet bowl outlet at one end, a drainpipe or soil stack at other end.

Code. Legal requirements for a plumbing installation.

Compression fitting. Easy-to-use connector for copper or plastic pipe.

Coupling. Fitting used to connect two lengths of pipe in a straight run.

CPVC (chlorinated polyvinyl chloride). Plastic pipe for hot water.

Critical distance. Maximum horizontal distance allowed between a fixture trap and a vent or soil stack.

Cross connection. Accidental plumbing connection that mixes contaminated water with potable water supply.

Drum trap. Trap shaped like a cylindrical drum with inlet and outlet at different levels; occasionally used for tubs and showers instead of curved pipe sections.

DWV (drain-waste and vent). System that carries away waste water and solid waste, allows sewer gases to escape, and maintains atmospheric pressure in drainpipes.

Elbow. Fitting used for making turns in pipe runs (for example, a 90-degree elbow makes a right-angle turn). A street elbow has one male and one female end.

Emitter. Water distribution device used in drip irrigation.

Escutcheon. Decorative piece that fits over a faucet body or pipe extending from a wall.

Female. Pipes, valves, or fittings with internal threads.

Fitting. Device used to join pipes.

Fixture unit. Equal to 7½ gallons or 1 cubic foot of waste water per minute.

Flange. Flat fitting or integral edging with holes to permit bolting together (a toilet bowl is bolted to a floor flange) or fastening to another surface (a tub is fastened to wall through an integral flange).

Flapper. Device that replaces a tank stopper in a toilet.

Flare fitting. Threaded fitting used on copper and plastic pipe that requires flaring one end of pipe.

Flexible connector. Bendable piece of pipe that delivers water from a shutoff valve to a fixture or appliance.

Float arm. Wire arm that connects float ball at one end to ball cock assembly at other end of a toilet tank.

Float ball. Large copper or plastic ball that floats on surface of water in a toilet tank and that descends and ascends with water level.

Floor flange. Fitting that connects a toilet to floor and drainpipe.

Flue. Large pipe through which fumes escape from a gas water heater.

Fluorocarbon tape. Special tape used as a joint sealer in place of pipe joint compound. Also called pipe-wrap tape.

Flush valve. Valve comprised of a stopper and valve seat that controls flow of water from toilet tank into bowl.

Flux. Paste applied to copper pipe before soldering; prevents oxidation when heat is applied to metal.

Gasket. Device (usually rubber) used to make joint between two valve parts watertight.

Gate valve. Valve with a tapered member at end of stem; acts as a gate to control flow of water.

Globe valve. Valve with a washer at end of stem that fits into valve seat to stop flow of water.

Graphite packing. Wirelike material wrapped around a faucet stem to prevent leaking.

Gray water. Household water recycled from showers, bathtubs, sinks, and washing machines and used for watering some landscaping.

Hose bibb. Valve with an external threaded outlet for accepting a hose fitting.

House trap. U-shaped fitting with two adjacent cleanout plugs, visible at floor level if main drain runs under floor.

Hub. Cast iron DWV pipe with one bell-shaped end called a hub or bell and one lipped end called a spigot. Spigot end of one pipe fits into hub of another and is caulked with oakum or sealed with molten lead or cold-lead wool.

Hubless. Cast iron pipe joined with neoprene gaskets and clamps, making it much easier to use than hub pipe.

Indirect venting. Draining and venting a fixture into a floor drain or laundry tub.

Individual venting. Venting a fixture or fixture group individually to roof.

J bend. Piece of drainpipe in shape of letter J used with an elbow in a trap.

Joint. Point at which two sections of pipe are fitted together.

Locknut. Nut fitted onto one piece of pipe and screwed onto another piece to join two pieces.

Main cleanout. Fitting in shape of letter T near bottom of soil stack or where drain leaves house.

Male. Pipes, valves, or fittings with external threads.

Nipple. Short length (12 inches or less) of galvanized pipe with external threads on both ends for joining fittings.

No-hub. *See* Hubless

Oakum. Stranded hemp used in making hub (bell and spigot) joints of cast iron pipe watertight.

O-ring. Narrow rubber ring; used in some faucets instead of packing to prevent leaking around stem and in swivel-spout faucets to prevent leaking at base of spout.

Packing. *See* Graphite packing

Packing nut. Nut screwed down onto stem of a faucet, holding packing tight.

PB (polybutylene). Flexible plastic tubing for hot or cold water.

PE (polyethylene). Flexible plastic tubing for cold water.

Penetrating oil. Used to help loosen a threaded joint in which corrosion has fused fittings.

Pipe cutter. Tool designed for making perfectly square cuts in a pipe.

Pipe joint compound. Sealing compound used on threaded fittings (applied to external threads).

Pipe-wrap tape. *See* Fluorocarbon tape

Plug. Closed-end, externally threaded fitting for closing off a pipe end that has internal threads.

Power snake. Electrically powered snake used for clearing blockages in main drain or house sewer. Available at rental stores.

PP (polypropylene). Rigid plastic pipe used for traps and drainpipes.

Pressure regulator. Device installed on a water supply line to reduce water pressure.

P trap. Fixture trap in shape of letter P.

PVC (polyvinyl chloride). Rigid plastic pipe for cold water.

Reamer. Tool that fits into pipe ends and is used to grind off internal burrs caused by cutting pipe. Often sold in combination with a pipe cutter.

Reducer. Fitting that connects pipe of one diameter with pipe of a smaller diameter.

Riser. Vertical run of pipe.

Run. Horizontal or vertical series of pipes.

Saddle tee (or T). Fitting for copper or galvanized pipe that is bolted onto pipe, eliminating cutting and threading or soldering; usually requires drilling into pipe.

Sanitary fitting. Fitting with no inside shoulders to block flow of waste; used to join DWV pipe.

Seat. Valve part into which washer or other piece fits, stopping flow of water.

Secondary venting. Venting fixtures distant from main stack to roof through a second vent.

Siphoning. Action occurring when a vacuum in a pipe pulls nearby water into it.

Slip fitting. Copper or plastic coupling without an interior ridge or shoulder used when repairing or extending pipe.

Sludge. Solid waste matter that settles to bottom of a septic tank.

Snake. Springlike tool forced into waste lines to break up blockages. Also called an auger.

Soil stack. Large DWV pipe that connects toilet and branch drains to house drain and also extends up and out house roof; upper portion serves as a vent.

Solder. Soft metal wire used to bond copper pipes.

Solvent cement. Compound used to join rigid plastic pipe.

Spacers. Short pieces of plastic or copper pipe cut to size; used when repairing or extending pipe.

Standpipe. Special drainpipe for a washing machine.

Stopper. Device that fits over flush valve opening at bottom of a toilet tank. When raised, permits water in tank to flow downward into bowl.

Stubout. End of a supply pipe or drainpipe that extends from wall or floor.

Sweating. Name for soldering joints. Also, an accumulation of moisture on pipes and tanks caused by condensation when cooler surface of pipe or tank meets warm air.

Swivel-head washer. Washer with a swivel base fixed to a clip top.

Tank ball. *See* Flapper

T (or tee). T-shaped fitting with three openings.

Temperature and pressure relief valve. Safety valve for water heater that lets water and steam escape.

Tempering valve. Valve that mixes a small amount of hot water with cold water entering a toilet tank to prevent sweating.

Threader. Tool used for cutting threads into pipe.

Toggle bolt. Bolt with two hinged wings used to fasten brackets to wall materials; can also be used to repair leaks in water tanks.

Torch. Tool that uses acetylene, butane, or propane to solder pipes together.

Transition fitting. Fitting that joins pipes of different materials. Also called an adapter.

Trap. Device (most often a curved section of pipe) that holds a water seal to prevent sewer gases from escaping into a home through a fixture drain.

ULF toilet. Ultra-low-flush toilet that uses no more than 1.6 gallons of water per flush.

Union. Fitting that joins two lengths of pipe and permits assembly and disassembly without taking entire section apart.

Valve. Device that controls flow of water.

Valve seat. *See* Seat

Valve seat dresser. Tool used to grind down burrs on a valve seat.

Valve seat wrench. Hexagonal-end wrench inserted into hexagonal opening in a valve seat for installing or removing seat.

Water hammer. Sound of pipes shuddering and banging.

Wax gasket. Donut-shaped seal, made of wax, used at base of a toilet to prevent drainpipe leaks.

Wet venting. Venting arrangement in which a fixture's drainpipe, tied directly to soil stack, vents fixture also.

Y (or wye). Fitting with three outlets in shape of letter Y.

Index

Proof-of-Purchase
ISBN 0-376-01467-9 (SB)
ISBN 0-376-09037-5 (SL)